Geophysical Monograph Series

Including

IUGG Volumes
Maurice Ewing Volumes
Mineral Physics Volumes

GEOPHYSICAL MONOGRAPH SERIES

Geophysical Monograph Volumes

1. Antarctica in the International Geophysical Year *A. P. Crary, L. M. Gould, E. O. Hulburt, Hugh Odishaw, and Waldo E. Smith (Eds.)*
2. Geophysics and the IGY *Hugh Odishaw and Stanley Ruttenberg (Eds.)*
3. Atmospheric Chemistry of Chlorine and Sulfur Compounds *James P. Lodge, Jr. (Ed.)*
4. Contemporary Geodesy *Charles A. Whitten and Kenneth H. Drummond (Eds.)*
5. Physics of Precipitation *Helmut Weickmann (Ed.)*
6. The Crust of the Pacific Basin *Gordon A. Macdonald and Hisashi Kuno (Eds.)*
7. Antarctica Research: The Matthew Fontaine Maury Memorial Symposium *H. Wexler, M. J. Rubin, and J. E. Caskey, Jr. (Eds.)*
8. Terrestrial Heat Flow *William H. K. Lee (Ed.)*
9. Gravity Anomalies: Unsurveyed Areas *Hyman Orlin (Ed.)*
10. The Earth Beneath the Continents: A Volume of Geophysical Studies in Honor of Merle A. Tuve *John S. Steinhart and T. Jefferson Smith (Eds.)*
11. Isotope Techniques in the Hydrologic Cycle *Glenn E. Stout (Ed.)*
12. The Crust and Upper Mantle of the Pacific Area *Leon Knopoff, Charles L. Drake, and Pembroke J. Hart (Eds.)*
13. The Earth's Crust and Upper Mantle *Pembroke J. Hart (Ed.)*
14. The Structure and Physical Properties of the Earth's Crust *John G. Heacock (Ed.)*
15. The Use of Artificial Satellites for Geodesy *Soren W. Henricksen, Armando Mancini, and Bernard H. Chovitz (Eds.)*
16. Flow and Fracture of Rocks *H. C. Heard, I. Y. Borg, N. L. Carter, and C. B. Raleigh (Eds.)*
17. Man-Made Lakes: Their Problems and Environmental Effects *William C. Ackermann, Gilbert F. White, and E. B. Worthington (Eds.)*
18. The Upper Atmosphere in Motion: A Selection of Papers With Annotation *C. O. Hines and Colleagues*
19. The Geophysics of the Pacific Ocean Basin and Its Margin: A Volume in Honor of George P. Woollard *George H. Sutton, Murli H. Manghnani, and Ralph Moberly (Eds.)*
20. The Earth's Crust: Its Nature and Physical Properties *John G. Heacock (Ed.)*
21. Quantitative Modeling of Magnetospheric Processes *W. P. Olson (Ed.)*
22. Derivation, Meaning, and Use of Geomagnetic Indices *P. N. Mayaud*
23. The Tectonic and Geologic Evolution of Southeast Asian Seas and Islands *Dennis E. Hayes (Ed.)*
24. Mechanical Behavior of Crustal Rocks: The Handin Volume *N. L. Carter, M. Friedman, J. M. Logan, and D. W. Stearns (Eds.)*
25. Physics of Auroral Arc Formation *S.-I. Akasofu and J. R. Kan (Eds.)*
26. Heterogeneous Atmospheric Chemistry *David R. Schryer (Ed.)*
27. The Tectonic and Geologic Evolution of Southeast Asian Seas and Islands: Part 2 *Dennis E. Hayes (Ed.)*
28. Magnetospheric Currents *Thomas A. Potemra (Ed.)*
29. Climate Processes and Climate Sensitivity (Maurice Ewing Volume 5) *James E. Hansen and Taro Takahashi (Eds.)*
30. Magnetic Reconnection in Space and Laboratory Plasmas *Edward W. Hones, Jr. (Ed.)*
31. Point Defects in Minerals (Mineral Physics Volume 1) *Robert N. Schock (Ed.)*
32. The Carbon Cycle and Atmospheric CO_2: Natural Variations Archean to Present *E. T. Sundquist and W. S. Broecker (Eds.)*
33. Greenland Ice Core: Geophysics, Geochemistry, and the Environment *C. C. Langway, Jr., H. Oeschger, and W. Dansgaard (Eds.).*
34. Collisionless Shocks in the Heliosphere: A Tutorial Review *Robert G. Stone and Bruce T. Tsurutani (Eds.)*
35. Collisionless Shocks in the Heliosphere: Reviews of Current Research *Bruce T. Tsurutani and Robert G. Stone (Eds.)*
36. Mineral and Rock Deformation: Laboratory Studies—The Paterson Volume *B. E. Hobbs and H. C. Heard (Eds.)*
37. Earthquake Source Mechanics (Maurice Ewing Volume 6) *Shamita Das, John Boatwright, and Christopher H. Scholz (Eds.)*

38 **Ion Acceleration in the Magnetosphere and Ionosphere** *Tom Chang (Ed.)*
39 **High Pressure Research in Mineral Physics (Mineral Physics Volume 2)** *Murli H. Manghnani and Yasuhiko Syono (Eds.)*
40 **Gondwana Six: Structure, Tectonics, and Geophysics** *Garry D. McKenzie (Ed.)*
41 **Gondwana Six: Stratigraphy, Sedimentology, and Paleontology** *Garry D. McKenzie (Ed.)*
42 **Flow and Transport Through Unsaturated Fractured Rock** *Daniel D. Evans and Thomas J. Nicholson (Eds.)*
43 **Seamounts, Islands, and Atolls** *Barbara H. Keating, Patricia Fryer, Rodey Batiza, and George W. Boehlert (Eds.)*
44 **Modeling Magnetospheric Plasma** *T. E. Moore and J. H. Waite, Jr. (Eds.)*
45 **Perovskite: A Structure of Great Interest to Geophysics and Materials Science** *Alexandra Navrotsky and Donald J. Weidner (Eds.)*
46 **Structure and Dynamics of Earth's Deep Interior (IUGG Volume 1)** *D. E. Smylie and Raymond Hide (Eds.)*
47 **Hydrological Regimes and Their Subsurface Thermal Effects (IUGG Volume 2)** *Alan E. Beck, Grant Garven, and Lajos Stegena (Eds.)*
48 **Origin and Evolution of Sedimentary Basins and Their Energy and Mineral Resources (IUGG Volume 3)** *Raymond A. Price (Ed.)*
49 **Slow Deformation and Transmission of Stress in the Earth (IUGG Volume 4)** *Steven C. Cohen and Petr Vaníček (Eds.)*
50 **Deep Structure and Past Kinematics of Accreted Terranes (IUGG Volume 5)** *John W. Hillhouse (Ed.)*
51 **Properties and Processes of Earth's Lower Crust (IUGG Volume 6)** *Robert F. Mereu, Stephan Mueller, and David M. Fountain (Eds.)*
52 **Understanding Climate Change (IUGG Volume 7)** *Andre L. Berger, Robert E. Dickinson, and J. Kidson (Eds.)*
53 **Plasma Waves and Istabilities at Comets and in Magnetospheres** *Bruce T. Tsurutani and Hiroshi Oya (Eds.)*
54 **Solar System Plasma Physics** *J. H. Waite, Jr., J. L. Burch, and R. L. Moore (Eds.)*
55 **Aspects of Climate Variability in the Pacific and Western Americas** *David H. Peterson (Ed.)*
56 **The Brittle-Ductile Transition in Rocks** *A. G. Duba, W. B. Durham, J. W. Handin, and H. F. Wang (Eds.)*
57 **Evolution of Mid Ocean Ridges (IUGG Volume 8)** *John M. Sinton (Ed.)*
58 **Physics of Magnetic Flux Ropes** *C. T. Russell, E. R. Priest, and L. C. Lee (Eds.)*
59 **Variations in Earth Rotation (IUGG Volume 6)** *Dennis D. McCarthy and William E. Carter (Eds.)*

Maurice Ewing Volumes

1 **Island Arcs, Deep Sea Trenches, and Back-Arc Basins** *Manik Talwani and Walter C. Pitman III (Eds.)*
2 **Deep Drilling Results in the Atlantic Ocean: Ocean Crust** *Manik Talwani, Christopher G. Harrison, and Dennis E. Hayes (Eds.)*
3 **Deep Drilling Results in the Atlantic Ocean: Continental Margins and Paleoenvironment** *Manik Talwani, William Hay, and William B. F. Ryan (Eds.)*
4 **Earthquake Prediction—An International Review** *David W. Simpson and Paul G. Richards (Eds.)*
5 **Climate Processes and Climate Sensitivity** *James E. Hansen and Taro Takahashi (Eds.)*
6 **Earthquake Source Mechanics** *Shamita Das, John Boatwright, and Christopher H. Scholz (Eds.)*

IUGG Volumes

1 **Structure and Dynamics of Earth's Deep Interior** *D. E. Smylie and Raymond Hide (Eds.)*
2 **Hydrological Regimes and Their Subsurface Thermal Effects** *Alan E. Beck, Grant Garven, and Lajos Stegena (Eds.)*
3 **Origin and Evolution of Sedimentary Basins and Their Energy and Mineral Resources** *Raymond A. Price (Ed.)*
4 **Slow Deformation and Transmission of Stress in the Earth** *Steven C. Cohen and Petr Vaníček (Eds.)*
5 **Deep Structure and Past Kinematics of Accreted Terranes** *John W. Hillhouse (Ed.)*
6 **Properties and Processes of Earth's Lower Crust** *Robert F. Mereu, Stephan Mueller, and David M. Fountain (Eds.)*
7 **Understanding Climate Change** *Andre L. Berger, Robert E. Dickinson, and J. Kidson (Eds.)*
8 **Evolution of Mid Ocean Ridges** *John M. Sinton (Ed.)*
9 **Variations in Earth Rotation** *Dennis D. McCarthy and William E. Carter (Eds.)*

Mineral Physics Volumes

1 **Point Defects in Minerals** *Robert N. Schock (Ed.)*
2 **High Pressure Research in Mineral Physics** *Murli H. Manghnani and Yasuhiko Syono (Eds.)*

Geophysical Monograph 60
IUGG Volume 10

Quo Vadimus
Geophysics for the Next Generation

George D. Garland
John R. Apel
Editors

Ⓢ American Geophysical Union
⊕ International Union of Geodesy and Geophysics

Geophysical Monograph/IUGG Series

Library of Congress Cataloging-in-Publication Data

Quo vadimus : geophysics for the next generation / George D. Garland, John R. Apel, editors.
    p. cm. — (Geophysical monograph ; 60) (IUGG ; v. 10)
  ISBN 0-087590-455-6
    1. Geophysics.   2. Geodesy.   I. Garland, George D. (George David), 1926– . II. Apel, John R. III. American Geophysical Union. IV. International Union of Geodesy and Geophysics.
V. Series. VI. Series: IUGG (Series) ; v. 10.
QC801.Q6 1990
550—dc20   90-23673
                                                                                                          CIP

ISBN 0-87590-455-6

Copyright 1990 by the American Geophysical Union, 2000 Florida Avenue, NW, Washington, DC 20009

Figures, tables, and short excerpts may be reprinted in scientific books and journals if the source is properly cited.

    Authorization to photocopy items for internal or personal use, or the internal or personal use of specific clients, is granted by the American Geophysical Union for libraries and other users registered with the Copyright Clearance Center (CCC) Transactional Reporting Service, provided that the base fee of $1.00 per copy plus $0.10 per page is paid directly to CCC, 21 Congress Street, Salem, MA 10970. 0065-8448/89/S01. + .10.
    This consent does not extend to other kinds of copying, such as copying for creating new collective works or for resale. The reproduction of multiple copies and the use of full articles or the use of extracts, including figures and tables, for commercial purposes requires permission from AGU.

Printed in the United States of America.

CONTENTS

Preface
George D. Garland and John R. Apel ix

Vening Meinesz, a Pioneer in Earth Sciences
N. J. Vlaar xi

Geodesy and Geophysics in Their Interaction with Mathematics and Physics, and Some Open Problems in Geodesy
Helmut Moritz 1

Geodesy and Geophysics
J. M. Wahr 5

Geophysical Geodesy: The Study of the Slow Deformations of the Earth
Kurt Lambeck 7

Some Possible Additional Answers
Petr Vaníček 11

The Earth as a Planet
William M. Kaula 13

Space Plasma Physics
Donald J. Williams, George L. Siscoe 21

The Importance of the Variability of the Solar-Terrestrial Environment
Y. Kamide 23

Where Are We Going in the Study of Short-Period Climate Fluctutations?
Jerome Namias 25

Geophysical Fluid Dynamics and Related Topics
Raymond Hide 39

Hydrodynamic Complexity in the Earth System
W. Richard Peltier 43

Climate
George C. Reid 45

Comments on George Reid's "Quo Vadimus" Contribution "Climate"
Paul J. Crutzen 47

Research in Climate Science
Stephen H. Schneider 49

Scale Interactions
John T. Snow 55

Oceanography and Geophysics
D. Lal 59

Contents

Earth Science in the 21st Century
Paul H. LeBlond   63

Physical Oceanography to the End of the Twentieth Century
R. W. Stewart   65

Comments on R. W. Stewart's "Physical Oceanography to the End of the Twentieth Century"
C. Wunsch   69

Quo Vadimus—Hydrology
Zbigniew W. Kundzewicz   71

Statement to Follow Quo Vadimus—Hydrology by Z. W. Kundzewicz
Mark Meier   77

Statement on Quo Vadimus—Hydrology
Peter S. Eagleson   79

The Lithosphere of the Earth as a Large Non-Linear System
V. Keilis-Borok   81

Contributions to the Quo Vadimus Symposium   83

Index   115

PREFACE

This Monograph contains the contributions to the Union symposium, held at the XIX General Assembly, Vancouver, Canada in August 1987, and known as the Quo Vadimus symposium.

As the name ("Where are we going?") suggests, one principal purpose of the symposium was to identify outstanding problems of geodesy and geophysics, and thereby to stimulate efforts to solve them over the next few decades. While there had been a number of proposals to take a critical look at the present state of these sciences, the form of the symposium owes much to a suggestion by Professor Keilis-Borok, that geodesists and geophysicists emulate the critical review of mathematics made by David Hilbert in 1900. Hilbert outlined twenty-three unsolved problems, which he felt were being bequeathed to the mathematicians of the twentieth century. His paper has had great influence, and many of the problems isolated by him have now been solved.

There was a second major purpose of the symposium: to mark the centenary of the birth of the great Dutch geodesist and geophysicist F.A. Vening Meinesz (born 1887) and to honor his memory. It was felt that this was particularly appropriate, since Vening Meinesz, who was President of the Union from 1948 to 1951, would himself have been fascinated by the problems and by the approach taken to identify them.

The arrangement of this Monograph is based in part upon the format of the symposium. Well in advance of the 1987 meeting, invitations were sent by the symposium committee (Professors V.I. Keilis-Borok, Helmut Moritz and G.D. Garland) through all National Committees and Associations, inviting brief individual statements of perceived major problems. These statements were collected and printed by Professor Moritz and made available at the symposium. They appear in the second part of this Monograph, without further editing or reviewing.

The papers presented at the symposium were those of speakers invited by the symposium committee. These papers form the nucleus of the first part of the Monograph. In order to prepare the material for publication, the AGU appointed Professor G.D. Garland and Dr. J.R. Apel as volume editors. In the course of arranging the written contributions for peer review, the editors noted that there were aspects of geodesy and geophysics which had not been considered by any speaker. To remedy this, the editors invited additional written communications, and also invited all reviewers of the papers, if they wished, to prepare their comments for publication.

We believe that the volume now presented gives a reasonably balanced view of the outstanding problems of geodesy and geophysics. The authors vary, of course, in the optimism with which they consider the solutions to these problems to be forthcoming, some writers emphasizing the difficulties, others the advances which are already being made. Let us hope that this critical look at our sciences will bear fruit as did Hilbert's analysis of the state of mathematics in 1900.

George D. Garland
Department of Physics
University of Toronto

John R. Apel
Applied Physics Laboratory
Johns Hopkins University

F. A. Vening Meinesz

VENING MEINESZ, A PIONEER IN EARTH SCIENCES

N.J. Vlaar
Vening Meinesz Laboratories
University of Utrecht
Budapestlaan 4
3584 CD Utrecht
The Netherlands

F.A. Vening Meinesz (1887-1966) is especially known from his gravity measurements at sea. He devised the Vening Meinesz pendulum apparatus by which it became possible to measure gravity at sea with comparable accuracy as on land.

Starting in 1923 he conducted several global gravity surveys on voyages on submarines, particularly to and in the Indonesian Archipelago. He detected strong gravity anomaly belts running parallel to the Indonesian deep sea trenches. He explained these Meinesz belts as sites of downbuckling of the Earth's crust. He introduced the concept of regional isostasy taking flexure of an elastic crust into account. He also contributed to physical geodesy: The Vening Meinesz formula connects the deviation of the vertical from the plumbline to gravity anomalies.

At the beginnings of space age the problem of the Earth's external gravity field received his attention.

His work connected with geodynamics, and in particular with mountain building, epeirogeny, and graben tectonics, did not receive much of the interest it certainly deserves. The present paper deals in particular with the geodynamic concepts which were developed by Vening Meinesz as a consequence of his observational work, already from an early time onward. The evolution of these concepts, from the possibility of a contracting Earth to the hypothesis of convection currents in the Earth as underlying causes of mountain building and other geodynamic processes, are presented in retrospect. Special attention is devoted to the connection between Vening Meinesz' ideas and their relevance for the plate tectonic hypothesis.

Vening Meinesz stood at the threshold of modern Earth science. Many of his ideas have not yet lost their validity. Many problems he addressed still await their solution. The social and scientific environment in which he worked enabled individuals like him to pioneer pathways into the future.

Measuring Gravity at Sea

Felix Andries Vening Meinesz was born on July 30, 1887, as the youngest of the four children of S.A. Vening Meinesz and C.A.C. den Tex. He descended from lines of Dutch gentry and patriciate. His father was mayor of the city of Rotterdam and later of Amsterdam. His background, stately appearance, and distinguished manners would, in his later scientific career, open many doors and give access to the highest authorities.

The choice of his high school education indicates that he destined himself for a profession in science, technology, or commerce. This was quite unusual in his family. A career as magistrate or lawyer would have been more in line.

He graduated from Delft Technical University in civil engineering in 1910 and found his first occupation with the Netherlands State Committee on Arc measurements. He was assigned to measure gravity in the Netherlands. One of the main concerns of the international geodetic community at the beginning of this century was to determine the Earth's shape, not only by arc measurements, but also by means of the gravity field. Measuring the gravity field in the Netherlands would be decisive for Vening Meinesz' future career.

Like for the other famous Dutch geodesist, Snellius, some centuries before, the peculiar geography of the low countries would prove to be the incentive for some of their most important contributions to science. The flat landscape, together with the many church steeples to act as reference points, induced Snellius to become the founder of geodetic triangulation. Vening Meinesz designed his famous pendulum apparatus in order to eliminate the noise of the unstable and shaky soil. The principle of the apparatus was based on using two pendulums swinging in a common plane with equal amplitudes and opposite phases. Thus their relative motion eliminates the motion of their common support. Later, he perfected his instrument by adding a third pendulum.

The theory of his apparatus was subject of the thesis, which, in 1915, earned him the doctors degree, cum laude. Up to 1921 gravity in the Netherlands was measured at fifty localities. The precision of the measurements was shown to be considerably increased relative to single pendulum observations and this success induced Vening Meinesz to investigate the feasibility to use the instrument at sea. Gravity at sea, which covered the larger part of the Earth's surface, was not yet possible with the same precision as on land. Earlier measurements at sea by Hecker of the Berlin observatory, who used an instrument based on barometric principles, did not achieve a precision comparable to on land measurements. He succeeded in finding large gravity variations over the Tonga-Kermadec trench.

Vening Meinesz had to employ his apparatus on board a submarine in order to eliminate as much as possible unwanted wave motion. In 1923 he took his instrument on a voyage of one of the submarines of the Dutch Navy, H.Ms.K2, sailing from the Netherlands to Indonesia -- the former Dutch East Indies -- through the Suez Canal. Several submarines had to be brought to the East Indies in order to strengthen the Dutch fleet in the area. Vening Meinesz was granted to exploit this opportunity for scientific purposes. His primary goal was to test his instrument. After some initial misfortune he arrived at measuring absolute gravity with an accuracy of the order of 1 mgal, hence, at the epoch, comparable to measurements at land.

Several voyages aboard submarines to the far East were to follow. In 1926 with H.Ms.K13 through the Panama Canal, and in 1935 around the Cape of Good Hope. In 1937, Vening Meinesz participated in a voyage to the United States aboard the submarine H.Ms.O16. He also took part, taking his pendulum apparatus along, in submarines of the U.S. and of the Italian Navy, for the purpose of studying the gravity field in the Caribbean and the Mediterranean seas, respectively.

Being of tall stature, about two meters tall, life on those early small submarines must have meant hardship for Vening Meinesz. Adding to this the heat, either of the tropical sun when at the surface, or of the batteries which powered the ship under water, great sacrifices had to be made during the more than 100,000 nautical miles totalling the voyages to the East Indies.

Though the primary purpose was to take the submarines to the overseas territories, the scientific goals also were considered to be of great importance. After arrival in Indonesia, the K13 was put to Vening Meinesz' disposal for half a year, solely for the purpose of measuring bathymetry and gravity in the Indonesian waters. The cooperativeness of the Dutch admiralty was set as an example by the international scientific community to other seafaring nations. Of course, showing the flag, and trying to place orders for submarines in overseas countries also played their role. The combination of naval and scientific achievements came in as an extra argument. In that sense Vening Meinesz could play a part in public relations! The O16 trip to the U.S.A. in 1937 even gave rise to being received by President Roosevelt.

The unescorted voyage of the K13 in 1926 of over 20,000 nautical miles was the largest ever made by a submarine and resulted in a book written by two officers of the crew.

The 1930's were a time of records, and the combination of science, technology and travels to far away beaches stirred the imagination. In 1936, the voyage of the K17 was made subject of a movie picture, featuring the crew and Vening Meinesz. For the older generation in Holland, the K17 still rings a bell.

Vening Meinesz designed his instrument, to become known as the Vening Meinesz pendulum apparatus, and undertook personally to measure gravity at sea. Notwithstanding the elaborate measurements and corrections, hundreds of pendulum observations over the world oceans, and particularly in the Indonesian Archipelago were made. These achievements must be esteemed as highly outstanding contributions to geodesy and geophysics at that time. It would take another 35 years after its invention before the Vening Meinesz apparatus could be replaced by a more modern instrument for measuring gravity at sea with the same accuracy: the Askania spring gravimeter placed on a stable platform aboard a surface ship.

The Meinesz Belt and the Buckling Hypothesis

The first aim of measuring gravity at sea was to have more pertinent data in order to determine the shape of the Earth. A particular issue was whether the Earth's equator would deviate from the circle. From his observations, at equatorial latitudes circumpassing the globe, Vening Meinesz found no evidence for such a deviation. He found, however, that like on the continents, also on the oceans Airy isostasy prevailed, thus confirming the existence of a thin low-density crust which floated on a high-density yielding mantle. Knowing that the continental topography could give rise to deviations from isostatic equilibrium, he was intrigued by the gravity effects of the deep sea trenches in the Indonesian Archipelago, and gravity was measured more densely in this area. This led to the discovery of belts of strong negative isostatic anomalies running parallel to the trenches. These belts were to become known as the Meinesz (-) belts. He found that these belts were inherent to deep sea trenches in general. The negative anomaly belt could only be explained by assuming that the elastic crust was held down in the trench by external causes, maintaining deviation from isostatic equilibrium. A second, parallel belt of positive anomalies, discovered more seaward from the trench, could be explained in terms of the upward flexure of the elastic crust being pushed down at the trench. The theory of the flexure of a thin elastic plate gave a thickness of 35 km for the crust. This is still, at present, a valid figure for the thickness of the mechanically strong part of the oceanic lithosphere. Vening Meinesz indeed distinguished between the chemical crust, which in general was subject to Airy isostasy, and the "elastic" crust, which was defined rheologically. The flexure of the elastic crust made him to broaden the concept of isostasy. Whereas Airy isostasy departed from purely local density compensation at depth, Vening Meinesz introduced regional isostasy and compensation, taking into account the flexure of the elastic crust when loaded. Regional or Vening Meinesz isostasy was to become a useful concept in connection with mountain building and loading of the oceanic lithosphere by volcanic edifices.

As early as 1930, Vening Meinesz casts his findings in a more general framework. He was aware of a connection of the Meinesz belts, deep sea troughs, seismicity, volcanism, and mountain building. Realizing that strong isostatic anomalies could only be supported by forces acting on the elastic crust, he suggested that these forces, acting in the plane of crust, should give rise to linear features of local plastic yielding. This then should lead to local thickening and downbuckling of the crust. Relaxation of the acting force fields would restore isostatic equilibrium and uprising of the downbuckled crust would lead to the formation of a mountain belt. He had found that the Meinesz belts occurred generally on the Western Pacific rim. Being aware of the large forces involved and the large regional extent of their action, he speculated that downbuckling should be the result of compression between large, and more or less rigid, crustal blocks. In 1931 he regards contraction theory as most plausible generating cause for the large compression required, and also mentions Wegener, without, however, coming to conclusions concerning a causal mechanism. His views found support in the ideas living among contemporary Dutch geologists and engineers.

The geologist Umbgrove (The pulse of the Earth) proposed that mountain building periods were episodically occurring during the Earth's history. This confirmed Vening Meinesz' hypothesis that mountain building was the effect of forces on the Earth's crust, which, after a period of action, would relax. The sedimentologist Kuenen would later provide scale experiments by which the downbuckling process was vividly demonstrated.

In 1935, the engineer Bylaard published an article in which he associated the theory of plastic buckling of an elastic plate with the findings of Vening Meinesz in the East Indies. When an elastic plate is subject to increasing in-plate compression, instability will occur, giving rise to large displacements perpendicular to the plate, and along in-plate lines perpendicular to the direction of compression. However, if the plastic yield limit is reached before elastic instability, plastic thickening will take place along a line making an angle of 55° with the direction of compression. Applying this to the Earth's elastic crust should lead to thickening and downbuckling in the plastic substratum and thus to crustal shortening and the formation of a root as depicted in Kuenen's experiment.

Later, following David Griggs (1939), he also distinguished the case of plastic shear along a crustal fault, taking place without plastic thickening, and allowing movements such as along the San Andreas fault. Theory predicted that this should take place under an angle of 23° with the direction of compression. The angles of 55° and of 23°, in Vening Meinesz' later work, were almost to become magical. He found these angles clearly demonstrated along the troughs south of Java and southwest of Sumatra by assuming a pressure field which was aligned N.W.-S.E., comprising the Australian-Indian Ocean and the Asiatic block. Whereas south of Java plastic thickening and downbuckling was supposed to take place, Sumatra was to move laterally by "pseudo-viscous" shear with respect to the Indian Ocean. The latter shear movement, in a compressive regime, could cause overriding of one crustal block over the other and thus also could result in a form of mountain building. Indeed, already in 1898, J.J.A. Mulder, by geodetic triangulation near Bengkulen, Sumatra, had observed lateral off-sets of 2 meter along a fault directed parallel to the coast. Mulder, as president of the Dutch State Committee on Arc Measurements, would later be the first employer of Vening Meinesz.

The Convection Hypothesis

The first time that Vening Meinesz made mention of the possibility of convection currents in the Earth is in a highly remarkable paper which appeared in 1934 in the proceedings of the Royal Netherlands Academy of Science entitled: "Gravity and the Hypothesis of Convection Currents in the Earth." Referring to the hypothesis as being put forward by Arthur Holmes some years earlier for explaining the driving mechanism of mountain building, Vening Meinesz states that thermal convection may well be causally related. The motivation for accepting the hypothesis, however, is not *ad hoc*, but is based on clever reasoning departing from some peculiarities of the observed gravity field. He knew that large "fields" of anomalous gravity existed, both over continental areas and deep oceanic basins which up till then deferred being understood in terms of Airy isostasy. In order to explain the isostatic anomalies by anomalous density distributions, the Airy isostatic hypothesis, applied to the crust, did not give satisfactory answers, and subcrustal density anomalies had to be inferred.

Earlier, Vening Meinesz had suggested that upbending of the elastic crust by compression on a curved Earth would lead to subcrustal mass excess and thus to positive isostatic anomalies over deep sea and sedimentary basins. This suggestion fitted well in the hypothesis of a shrinking Earth. However, he found that the stresses involved should exceed 10 kilobar, and hence far beyond the yield limit of crustal materials.

Subcrustal mass excesses under deep ocean basins and mass deficiencies under large continental areas he now explained by temperature gradients from the continents to the ocean basins which by their inherent perturbation of hydrostatic equilibrium would give rise to subcrustal mantle currents from the continents to the oceans. These currents were then supposed to be part of a convective upper mantle circulation which was triggered by the horizontal temperature and associated horizontal pressure gradients. The temperature difference could be ascribed to blanketing by the continental sialic crust having a larger concentration of radioactive elements relative to the basaltic oceanic crust. A consequence of this scheme was episodicity of convective circulation which was in line with the ideas of geologists like Umbgrove, who advocated the idea of episodic mountain building. A further consequence was rapid uplift and subsidence as was evident in epeirogeny.

Simple calculations, using a viscosity as deduced from postglacial Fennoscandian uplift (in 1934!) confirmed, in a quantitative way, the picture of subcrustal upper mantle convection currents being episodically triggered by temperature gradients from the continents towards the oceans. Associated gravity anomalies were in good agreement. Convection currents were supposed to be rising under continents and sinking under oceans. The Earth's crust suffered drag at its base exerted by the subcrustal currents. When in a compressive regime the crust could be subject to downbuckling.

Later, during the forties, contrary to Bullard's hypothesis of compression, Vening Meinesz explained continental graben formation as being due to tension in the crust caused by upwelling and diverging hot material in a convective subcrustal regime.

Heat flow measurements in the ocean floor would only emerge after 1950. Had Vening Meinesz known modern oceanic heat flow data, his hypothesis would have predicted upwelling limbs of convection cells beneath the oceanic rises, and downwelling ones at the continent-ocean transition beneath the deep oceanic basins. His deduction of absence of Airy isostasy under the latter would be completely in accord with the modern concept of a spreading and cooling oceanic lithosphere. It would then have been one step more to involve the oceanic crust as part of the convective system. Subduction or ongoing downbuckling of oceanic lithosphere at deep sea trenches could have completed the cycle of creation and subduction of the oceanic crust as manifestation of thermal convection in the Earth. Lacking observational evidence on oceanic heat flow, Vening Meinesz had to assume that subcrustal currents were flowing from the continents towards the oceans, subjecting the Earth's crust to compression or tension.

Being aware of the strong deviations of hydrostaticity as expressed in gravity anomalies and most probably caused by horizontal temperature differences in the subcrustal "plastic layer", Vening Meinesz concluded that flow was a necessary consequence for readjustment of equilibrium. However, still in 1939 he states on the subject of the hypothesis of convection currents in the Earth: "For the time being it is nothing but a speculative hypothesis". His scepsis will turn into a belief, and even a faith in the hypothesis of thermal convection in the Earth, in his later years.

Already in the thirties it had been proved by Pekeris and by Hales that convection in a homogeneous layer with the thickness of the Earth's mantle, subjected to a slight superadiabatic gradient and with a viscosity deduced from postglacial rebound, produced a Rayleigh number which was far beyond the critical one.

Vening Meinesz adapted Rayleigh's theory to a spherical coordinate system and found that also in this case thermal convection throughout the mantle was feasible. His theory was intrinsically departing from the assumption of stationarity, notwithstanding that he took convection in the Earth as an intermittent process. He assumed that convective motion which was supposed to be triggered by horizontal temperature gradients, could only take place when gravitational instability of the Earth mantle from cooling from the Earth's surface downwards, was restored by overturning of the mantle. This should take place episodically in the Earth's history leading to periods of geosynclinal formation and mountain building, separated by long intervals of tectonic quiescence. We should now find ourselves in the decline of the Alpine period.

During the early forties Vening Meinesz used his theory to demonstrate that third order tesseral spherical harmonics would lead to the peculiar ocean-continent distribution which was known as the "tetraeder theory". Convection had to be mantle deep and was assumed to have taken place early in the Earth's history, having fixed the relative position of continents and oceans. Apart from localized downbuckling and rifting, the Earth's crust was supposed to remain a strong shell. Shrinking of the Earth was not anymore accepted, as graben structures proved to be tensional features in the crust. Wegener's theory of continental drift was only acceptable in as far as it explained the continent-ocean distribution being frozen-in already during the Earth's early existence. During the early period, the oceanic crust might have been still sufficiently weak by the prevailing high temperatures. Continents could drift in it, be swept together, or torn apart. Vening Meinesz rejected the notion of continental drift in more recent geological periods. Altogether, until the advent of the plate tectonic era, he stuck to this concept.

Following Prey (1922), who expanded the Earth's topography in spherical harmonics of lower order, Vening Meinesz extended the expansion to order 31. The spherical harmonic spectrum of the Earth's topography appeared to confirm his views that these features were caused during a period of thermal convection in the early stages of the Earth. In this context, also the oceanic rises were to be the frozen-in product of early geodynamical processes! However, admitting mantle convection to exert a net torque on the crust, he concluded that true polar wander would be well possible. He calculated the stress field in the thin crust generated when the latter shifts over the ellipsoidal mantle, and attributed large crustal fault systems to this mechanism. Later he will explain the results of paleomagnetism which indicate continental drift, in terms of true polar wander. Refinements of the concepts of thermal convection constitute the acceptance, apart from whole mantle convection, of smaller scale convection on an upper mantle scale. Whereas whole mantle convection appeared to be necessary for the explanation of large-scale compressive phenomena over the Pacific and South-East Asia, small-scale convection was required by the existence of small regional oceanic basins in Indonesia and the Mediterranean. It was known that these basins were rapidly subsiding.

During the fifties, inspired by the work of Meyering and Rooymans of Philips Physical Laboratories, and earlier suggestions by Bernal and others, he includes the olivine-spinel phase transition -- supposedly present in the upper mantle -- into his considerations. He concludes that this transition should enhance gravitational instability and thus whole mantle convection. Also, rapid basin subsidence could be fitted in the scheme of upper mantle phase changes. Whole mantle, or upper mantle convection, combined with downbuckling and shear failure of the elastic crust, enabled Vening Meinesz to set up a grand scenario for geodynamics. His approach should be characterized as speculative, but turns out to be a prelude to plate tectonics.

Mountain Building and Epeirogeny.

In geology it was observed that the initial stage in an orogenic cycle was the formation of a geosyncline. A geosyncline resulted from the subsidence of a sedimentary basin. After deposition of thick piles of sediments the sedimentary layering was folded and faulted. Subsequent uplift resulted in a mountain range above sea level.

Vening Meinesz regarded the orogenic cycle to be caused by plastic thickening, downbuckling and uplift of a narrow zone in a compressed elastic crust. The initial stage of plastic thickening and downbuckling resulted in a sagging depression in which sediments could be deposited. The unstable state of the crust resulted in accelerated downbuckling and collapse of the crustal layering. Whereas the lower part of the crust was to form a deep root penetrating the mantle, the more shallow crustal layering was compressed, faulted and folded.

This conception was in agreement with crustal shortening as observed in fold belts, for which the Alps stood as an example. Also, the existence of a "Verschluckungszone" (Swallowing zone) in the Alps as put forward by geologists was in agreement with the downbuckling hypothesis. A major difficulty, however, was posed by the amount of crustal material which had to be downbuckled into the mantle in order to be compatible with the amount of crustal shortening of the shallow crustal layering in the fold belt. The amount of downbuckled material, should, after restored isostasy as observed at present, have produced much higher mountain ranges. Vening Meinesz solved this problem by "subcrustal erosion". The light root material, after having been reheated to ambient mantle temperatures, would have been partly transported away by subcrustal currents. Adding this lighter material from below to the crust at distance, should then give rise to uplift of the crust and the formation of the German and French "Mittelgebirge".

Vening Meinesz distinguished between continental and oceanic geosynclines. He assumed that a mountain range like the Alps was the result of the collapse of a continental geosyncline. Due to the lack of a light sialic layer in the oceanic crust, mountain building should only be minor as only a small amount of light root would be available to produce large topographic features upon isostatic readjustment. In how far deep sea trenches at the margin of continental areas, like those in Indonesia, conform to this distinction, was not explicated. Island arc volcanism, however, was cast in terms of "subcrustal erosion" of a root containing more acid crustal material and producing magma.

A major difficulty, already in the first years of Vening Meinesz' involvement in Indonesian geodynamics, was posed by the presence of deep earthquakes in the Archipelago, which were detected by the seismologists Berlage, Vissers and Koning. Assuming subcrustal plasticity, the occurrence of these earthquakes could not be explained. The rheological model of "pseudo-plasticity" used by Vening Meinesz for the mantle, in which an initial small elastic strain had to be surpassed before plastic yielding occurred, enabled him to find an explanation. He argued that shallow earthquakes would trigger deep earthquakes when at depth the resulting deformation was too rapid to be relaxed by plastic flow.

Vening Meinesz distinguished three types of mountain ranges. Firstly, those associated with plastic downbuckling of the crust leading to Alpine type foldbelts. Secondly, mountain ranges which were supposed to result from overriding of one crustal block over the other on a "pseudo-viscous" shear fault in which the large pressure required for plastic downbuckling would not be reached. This mechanism of mountain building was in favor among American geophysicists. Its modern version is named transpression. Vening Meinesz attributed oceanic escarpments, which are now known to be associated with transform faults, to the latter mechanism. A third type of mountain ranges were constituted by the submarine oceanic ridge systems. At the time, he was not able to give a satisfactory explanation of their existence. However, by assuming that the ocean-continent distribution was generated in an early stage of the Earth's history, he speculated that they should be associated with true polar wander.

Plate Tectonics

During most of his speculative involvement with geodynamics, Vening Meinesz was only a few steps away from the hypothesis of plate tectonics. The phenomena he tried to explain were mostly the same as those which confirmed the spectacular breakthrough of plate tectonics. Vening Meinesz' model of the Earth's dynamics, however, appeared to be founded on so solid ground that he rejected the necessary notions inherent to plate tectonics.

As the Earth's crust and the continent-ocean distribution in his view originated in the beginnings of the existence of the planet, continental drift would not be possible in later periods. He disregarded geological evidence and also the findings of palaeomagnetism which were emerging from the early fifties onward. The mobility of the Earth's crust he understood as local yielding phenomena in an otherwise static crust. This "fixist" view of the Earth's crust was counterbalanced by assuming episodic convective overturning of the Earth's mantle, a "mobilistic" principle. Episodicity, however, excluded continuous relative motion of parts of the crust with respect to each other. Stationarity of the Earth's internal dynamics should have made ongoing downbuckling and thus creation of new crust necessary. In his episodic scenario there was no need for a crust which was part of the convective cycle. His notion that convection currents were upwelling under the continents and sinking under oceanic basins, was just the opposite from the findings which proved to be instrumental to the success of plate tectonics.

A few years before the beginnings of the plate tectonic era, influenced by the ideas of Hess and others, he embraced the notion that convection currents should be uprising under oceanic ridge systems as well as under continents.

His book, "The Earth's Crust and Mantle" appeared in 1964, just prior to the breakthrough of the plate tectonic hypothesis. The last two pages must be considered quite remarkable, particularly in the light of plate tectonics yet to come. He writes that Runcorn's palaeomagnetic findings are in complete accord with Wegener's theory and that continental drift therefore is well possible. He argues that downbuckling of the oceanic crust -- because of the small density difference between oceanic crust and mantle -- is easily to accomplish, and thus could go on indefinitely. This must be appreciated as the first statement concerning the possibility of subduction of oceanic crust. However, he fails to create necessary new oceanic crust elsewhere, otherwise plate tectonics possibly could have been born a few years earlier. Nevertheless, he still concludes that continental drift could only be possible during orogenies which he supposed to be episodic.

Vening Meinesz and Geology

In the early years Vening Meinesz found much appreciation for his ideas in geological circles. On the other hand, geology was very much in need of basic concepts which could explain mountain building, and he offered these concepts, firmly based on geophysical observations and theory. The geologists Umbgrove and Kuenen wrote articles on the geology of Indonesia in Vening Meinesz' "Gravity expeditions at sea, Part II" which appeared in 1934. However, this would be his last scientific cooperation with geologists.

Geologists amongst each other remained divided on tectonic issues for several decennia. Geology remained in a descriptive and naturalist state for many following years. Conflicting ideas and observations could only result in a chaotic presentation, not very much in accord with a schematic approach as advocated by Vening Meinesz. The distance between his approach and geological reality would remain too large to gain the sympathy of all geologists. As W. Nieuwenkamp once put it: "An extreme standpoint is to appreciate Vening Meinesz' work as an underhand attempt to torpedo the efforts of geologists from a submarine".

VENING MEINESZ, A PIONEER IN EARTH SCIENCES

Others understood the significance of geophysics at an early stage and encouraged Vening Meinesz to become a professor in geophysics -- on a part time basis -- at Utrecht University, already in 1937. His teachings would remain of limited significance as his courses had to cater to an audience with a poor background in mathematics and physics. Also, his approach stood far away from the ambitions of an average geology student who planned a career in industry. Studying the Earth as a physical system, even still at present, has not gained large popularity in many geological circles. Vening Meinesz always worked as an individualistic scientist. Also, geophysics had not evolved to the level of an indispensable tool in exploration and did not have the practical significance as it has at present. Probably all these circumstances together did not favor that a distinctive geophysical school around Vening Meinesz came into being. Vening Meinesz, being a professor in a geological department, showed his interest by participating in geological excursions. The students could not avoid the impression that professor Vening Meinesz admired the landscape and old village churches more than geology. On the other hand, the views on mountain building as developed by Vening Meinesz, particularly in connection with the geology of Indonesia and the Alps, were not always in agreement with geological interpretations.

The transport over large distances of nappes in fold belts like the Alps, was often accredited to gravitational sliding. This, of course, did not conform to the concept of the collapse of a zone of downbuckling.

Van Bemmelen, an authority on the geology of Indonesia, teaching at Utrecht University about contemporary with Vening Meinesz, held different opinions on the subject. Being an adept of Haarman's "undation theory" he thought to see this theory confirmed in the geology of Indonesia, demonstrating in his views an orogenetic wave travelling South East. Moreover, he ascribed orogeny in general, and nappe transport in particular, to gravitational sliding. Gravitational sliding was supposed to result from upbulging of the crust ("geoblisters", "geotumors"), due to upwelling of mantle material, which, in turn was mobilized by "physico-chemical processes". The latter process had the character of a "deus ex machina", not unlike present day "hot spots".

Van Bemmelen tried to proof his point in the field. Vening Meinesz could baffle his mostly geological audience with spherical harmonics and other mathematical intricacies. Discussions between those prominent scientists could run high. The atmosphere was not very encouraging for students to become dedicated to the problems of mountain building otherwise than by descriptive geological fieldwork.

Vening Meinesz was known to be a very amiable and courteous person. However, like many great men, he appeared to be blind for arguments which were not in line with his own views. He even could react with a certain intolerance towards unbelievers.

In the fifties and sixties, instigated by Martin Rutten, a professor in geology, the development of palaeomagnetism was strongly encouraged in Utrecht. Vening Meinesz never referred to this work. The idea of continental drift taking place in recent geologic times, did not appeal to him.

Epilogue

The first part of Vening Meinesz' career was marked by the great discoveries of his pendulum apparatus, the Meinesz belts, the concept of regional isostasy, and the downbuckling hypothesis. To this list we also should add the Vening Meinesz formula, connecting the deflection of the plumbline from the vertical to gravity anomalies. He devoted his later years to find a unifying theory for global tectonics. His lifework was ahead of his time. His background in civil engineering gave him the ability to regard the Earth as a mechanical system. Towards the end of his life he complained that he considered himself to be a failed civil engineer. He could permit himself this modesty as by then he was one of the greatest geophysicists to live. He truly was a great scientist.

Looking back, one feels great respect for his achievements and is impressed by the firmness and originality of his conclusions which were based on such a limited set of data. This then, compared to the present time, when data are drowning the world, science is managed by bandwagonry, and originality is often suppressed. Vening Meinesz was one of the last gentleman-scientists who were not bothered by collectivism and red tape.

In the meantime, plate tectonics has taught us a great deal about the evolution of the oceanic lithosphere, and insight has been gained in many geodynamical issues which were already addressed by Vening Meinesz.

Plate tectonics even appeared to confirm the convection hypothesis and has set the pace for a strong emphasis on the subject of convection in the Earth as an explanation for geodynamics. It may be that this emphasis may have reached the state of a paradigm, and even may hamper further developments. Plate tectonics yet did not contribute considerably to the subjects of Vening Meinesz' interest: mountain building and epeirogeny. Our state of knowledge of the continental evolution has hardly increased in a conceptual way since his days. Data and description have accumulated exponentially, but our understanding has not kept pace.

The subject of heat transfer in the interior of the Earth is interesting in its own right and of great importance for the evolution of our planet. However, this evolution cannot be understood properly without taking into account the growth and structure of the continental masses. Geophysicists should learn more from geological evidence, and geologists should be more aware of the possibilities and limitations of geophysics. The no man's land between the disciplines of geology, geophysics, and geodesy should be conquered by cooperation and not by domination.

Vening Meinesz stood high above his environment, physically, scientifically, and socially. He also stood out as a leader in scientific management: he was one of the founders of the Dutch counterpart of the National Science Foundation, and was General Director of the Royal Netherlands Institute at the Bilt. He held memberships of the Royal Netherlands Academy of Science, and also of several foreign Academies. He received many distinctions, both royal and academic. He was awarded the Bowie Medal in 1947, and the Vetlesen Prize in 1962. From 1933 to 1945 he was president of the International Association of Geodesy, and from 1948 to 1951 president of the International Union of Geodesy and Geophysics.

His life was devoted to science. He was a man with strong religious beliefs. Vening Meinesz died on August 10, 1966, leaving us a rich scientific heritage.

Quo Vadimus
Geophysics for the Next Generation

GEODESY AND GEOPHYSICS IN THEIR INTERACTION WITH MATHEMATICS AND PHYSICS, AND SOME OPEN PROBLEMS IN GEODESY

Helmut Moritz

Institute of Theoretical Geodesy,
Technical University, Graz, Austria

Abstract. The two questions, How can geodesy and geophysics benefit from mathematics and physics?, and conversely, How can mathematics and physics benefit from geodesy and geophysics? are posed and illustrated by some examples, reaching from Gauss' surface theory to the testing of gravitational theories. Then some concrete open problems in geodesy are formulated. Finally the question is raised whether the usual assumptions of smoothness or differentiability, e.g. concerning the earth's surface, are essential in view of the fact that features of the earth's surface have furnished some of the best illustrations of fractal theory.

Introduction

Geodetic and geophysical theory is clearly based on mathematics and physics. So we are users of these disciplines and benefiting from them. It is less known, however, that the opposite is also true: geodesy and geophysics have essentially contributed to the development of mathematics and physics in the past and may well continue to do so in the future, although to a more modest extent.

So we have two basic questions:
(A) How can geodesy and geophysics benefit from mathematics and physics?
(B) How can mathematics and physics benefit from geodesy and geophysics?

Question (A) is obvious and question (B) is relatively minor, but just because (B) is not so obvious, it may be worthwhile dwelling on it a little.

Some Examples

Certainly, the Quo Vadimus Symposium is directed to the future, but even in this endeavour we may learn from history. Thus, may I be permitted to mention some examples from the past as well as from the present.

Classical mathematics. The great French mathematicians of the 18th century (Clairaut, Laplace, Legendre, ...) have used questions regarding the figure of the earth to develop classical analysis and mathematical physics. Let me only mention spherical harmonics and potential theory, essentially contributing to the development of ordinary and partial differential equations.

Surface theory. Inspired by his practical geodetic work, Gauss discovered the intrinsic differential geometry of curved surfaces. Generalized to n dimensions by Riemann, this theory provided the mathematical basis for Einstein's General Theory of Relativity, which is considered the best presently available theory of gravitation.

Alternative gravitational theories. Many such theories have been proposed and experiments in the terrestrial environment can be used to test them. A good review of the enormous literature on the subject is found in the book [Misner et al., 1973, pp. 1047-1131]. Let me only mention what I consider the most spectacular example.

The most serious rival of General Relativity probably was a theory developed by the great logician and philosopher Alfred North Whitehead around 1920. Contrary to Einstein's theory, which is based on the differential geometry of a curved space-time, Whitehead's theory is the exact tensor analogue of Maxwell's vector equations for electromagnetism. It comes close to Einstein's theory in conceptual elegance and, furthermore, it has passed all the four standard tests of General Relativity (gravitational redshift, perihelion shift of Mercury, light deflection by the Sun, and radar time delay in the solar system). It even gives the same theory of black holes as Einstein's theory [Schild, 1962, p. 74]. Whitehead's theory is contradicted, however, by plain gravimetry: it predicts tidal variations of gravity of order $\Delta g/g \doteq 2 \times 10^{-7}$ with a 12 hour period which, needless to say, have never been observed [Misner et al., 1973, pp. 1067 and 1124].

The problem of testing gravitational theories

by gravimetry continues to be of interest, as the contribution of Treder in the present volume and the paper [Stacey et al., 1987] show. Keith Runcorn only recently expressed related ideas to me; and Hans Sünkel just called my attention to a pertinent brief notice in EOS, vol. 68, No. 35 (1987), p. 725.

These questions reach deep into the foundations of physics: unification of the gravitational force with the electromagnetic, the strong and the weak force of elementary particle physics, and perhaps the existence of a "fifth force". A similar fundamental problem of physics is the possible existence of magnetic monopoles, which might also be detected by geophysical methods, namely heat-flow measurements [Carrigan and Trower, 1982].

Fractals. So far, science has had an almost instinctive tendency to consider nature to be continuous and differentiable (this even holds for quantum theory). Curves that are continuous but not differentiable have, at best, been considered mathematical curiosities. The theory of fractals tries to show that such "monstrosities" can, after all, be quite useful in our description of nature. A glance at the fundamental book [Mandelbrot, 1983] shows the relevance of terrestrial phenomena such as coastlines, islands, lakes, snowflakes, and clouds. See also the contribution of Copaciu to this volume.

There are surprising cross-relations to the theory of dynamic systems (which was founded by Poincaré early in this century, inspired by a problem of planetary motion!) with its interplay between order and chaos; cf. Keilis-Borok's Union Lecture at the present General Assembly.

Some Open Problems in Geodesy

Traditionally, the task of geodesy has been the determination of the figure of the earth, of position of points on the earth's surface, and of its gravity field. In order for geodetic data to be useful for geophysics in the future (plate motion, variations in the earth's rotation, earth tides; see the paper by Wahr in this volume), a centimeter accuracy should be aimed at, both in absolute positioning and in the gravity field (a "centimeter geoid"). For positioning we have very precise methods such as Very Long Baseline Interferometry and satellite laser ranging, and the Global Positioning System (GPS) and similar systems promise an entirely new approach to geodetic positioning; see the contribution of Vanicek in this volume. A necessary prerequisite is the creation of a corresponding precise reference system; see the contributions of McCarthy and Wilkins.

Related problems are treated by Baran, Biró, Torge, Troitzkiy, Zieliński, and others in this volume, and the reader is invited to study their contributions in order to get a comprehensive picture. Let me give now my own list of favorite topics.

(a) Determination of the global fine structure of the gravity field, equivalent to a global geoid to an accuracy of a few centimeters. This would be important for many fields of geophysics, from crustal structure to ocean currents. Although this is one of the main tasks of geodesy, it must still be considered an open problem for technological, observational, theoretical, and political reasons.

(b) Finding density anomalies correlated to given gravity anomalies. The density anomalies in the earth determine the gravity anomalies uniquely, but not vice versa: the inverse problem is "improperly posed" in a mathematical sense. Still, finding, e.g., a set of possible density models consistent with a given gravity field and subject to certain "reasonable" boundary conditions would be important.

(c) Direct measurements of the geopotential. The classical way of determining geopotential differences is by levelling combined with gravity measurements, i.e., by a path integral of differential measurements, which is rather cumbersome and time-consuming. A direct measurement of the geopotential W is possible by the fact that the relativistic frequency shift of electromagnetic waves is proportional to W. The technological realization of this possibility will be a task of the future.

(d) Separation of gravity and inertia in moving measuring systems. Because of Einstein's equivalence of gravitation and inertia, both are (almost) inseparably related. Thus, gravity anomalies enter as noise in inertial surveying, and inertial disturbances occur as noise in aerial gravimetry. A separation of the two effects, say by combining accelerometers with gradiometers and/or GPS receivers, would make kinematical measurements in geodesy rigorous.

(e) Existence and uniqueness of the gravimetric boundary-value problem. Geodetic boundary value problems deal with the gravimetric determination of the earth's surface and its gravitational field. They are extremely difficult nonlinear free boundary-value problems of potential theory, of which only partial mathematical solutions have been obtained so far (for an unrealistically smooth earth surface); it is not even clear whether a complete solution can be achieved by contemporary mathematical tools.

(f) Determining a precise model for tidal variations in position. For the highly precise contemporary positioning methods of centimeter level, such as Very Long Baseline Interferometry or satellite laser ranging, a precise model (to an accuracy of 1 cm) of positional changes due to earth tides would be needed.

(g) <u>Modeling the irregular Chandler wobble</u>. Whereas precise mathematical models exist for forced (lunisolar) polar motion, free polar motion (Chandler wobble) has so far defied mathematical description because of its irregularity. The effects of atmospheric angular momentum appear to be particularly important for this purpose, but other little-understood geophysical effects, including the difficult problem of excitation of Chandler wobble, will be involved as well.

Let me finally make some general comments on this list.

Problems (c) and (d) are two examples of the increasing importance of General Relativity for geodesy and geophysics.

Concerning Problem (e), the gravimetric boundary-value problem, the state of the art can be seen from the book [Moritz, 1980], the review paper [Sansò, 1981] and, most recently, from the report [Grafarend et al., 1987], presented at this General Assembly. This problem is also addressed by Grafarend and Holota in the present volume.

As I said above, a complete and rigorous solution of this problem, for the very irregular earth surface, might even require completely new mathematical tools which do not yet exist. In the hands of a new Gauss or Poincaré, a solution of this problem could well help advance pure mathematics ...

Problems (b), (f) and (g) are only samples showing the interaction between contemporary and future geodesy and geophysics. Geodesy provides highly precise surface data useful for a geophysical study of the earth's interior; on the other hand, geodesy cannot be purely "superficial": it needs reference models for its data, which only solid-earth geophysics and meteorology can furnish.

"Nondifferentiable Geodesy"?

Geodesy, like geophysics and most physics in general, almost automatically works with "smooth" functions which are differentiable as often as required, apart from some "minor" discontinuities such as the core-mantle boundary. True, the external gravitational potential is a smooth, even an analytic function, but the earth's surface and the topographic masses very obviously are not smooth. This fact has been in the back of our minds all the time, but we have tried hard to forget it, approximating the earth's surface and the inertial mass distribution by smooth functions.

It is modern potential theory, with its concept of "measure" [cf. Anger, 1981] which has first opened our eyes that also extremely irregular mass distributions can be mathematically respectable. And it is rather embarassing to someone trained in the spirit of classical geodesy that Mandelbrot [1983] has taken the most impressive examples for his theory of fractals from features of the earth's surface, as we have mentioned above. It is even difficult to define exactly what the earth's surface is; cf. [Sansò, 1987].

I think it was Alfred North Whitehead who said that if a problem is too complicated to be solved by the usual methods, it might become solvable after having been suitably generalized. So, rather than "reducing" our data and "smoothing" the earth's surface, the problems of geodetic potential theory, including the boundary-value problem, might conceivably be tackled by taking the earth's surface and mass distribution seriously in their full realistic complexity and general irregularity. This sounds quite crazy, and I am myself rather sceptical about it. There is, however, at least one example when the geodetic boundary problem can be solved exactly for a very non-smooth boundary, namely if we consider as boundary the discrete set of observation points only, disregarding the rest of the earth's surface, cf. [Moritz, 1980, p. 95].

A famous physicist (I believe it was Niels Bohr) is reported to have said: "Your idea is crazy, but not crazy enough to be true". Let us hope that this does not apply here. At any rate, this problem is too eccentric to be included in the "respectable" list given above.

Concluding Remark

According to the topic of the present paper and the author's background, emphasis has been on problems of geodesy. It should be obvious, however, that geodesy and geophysics cannot be separated: they share common theoretical structures and practical problems, whose solutions will require an increasingly strong interrelation and cooperation. If there is a boundary between geodesy and geophysics, it certainly has a very "fractal" structure.

References

Anger, G., A characterization of the inverse gravimetric source problem through extremal measures, <u>Rev.Geophys. Space Phys.</u>, 19, 299-306, 1981 (reprint in [Grafarend and Rapp, 1984]).

Carrigan, R.A., and W.P. Trower, Superheavy magnetic monopoles, <u>Scientific American, 246(4)</u>, 91-99, 1982.

Grafarend, E.W., and R.H. Rapp (eds.), <u>Advances in Geodesy</u>, Selected Papers from Reviews of Geophysics and Space Physics, 310 pp., American Geophysical Union, Washington, D.C., 1984.

Grafarend, E.W., F. Sansò, and K.P. Schwarz (eds.), <u>Contributions to Geodetic Theory and</u>

Methodology, XIX General Assembly of IUGG, Publ. 60006, Dept. of Surveying Eng., Univ. of Calgary, 1987.

Mandelbrot, B.B., The Fractal Geometry of Nature, 468 pp., W.H. Freeman, New York, 1983.

Misner, C.W., K.S. Thorne, and J.A. Wheeler, Gravitation, 1279 pp., W.H. Freeman, San Francisco, 1973.

Moritz, H., Advanced Physical Geodesy, 500 pp., H. Wichmann, Karlsruhe, and Abacus Press, Tunbridge Wells, Kent, 1980.

Sansò, F., Recent advances in the theory of the geodetic boundary value problem, Rev. Geophys. Space Phys., 19, 437-449, 1981 (reprint in [Grafarend and Rapp, 1984]).

Sansò, F., Talk on the foundations of physical geodesy, in [Grafarend et al, 1987], pp. 5-27, 1987.

Schild, A., Gravitational theories of the Whitehead type and the principle of equivalence, in Evidence for Gravitational Theories, ed. by C. Møller, pp. 69-115, 1962.

Stacey, F.D., G.J. Tuck, and G.I. Moore, Geophysical observations relating to theories of quantum gravity, IUGG General Assembly, Vancouver, Abstracts V. 1, p. 205, 1987.

GEODESY AND GEOPHYSICS

J.M. Wahr

Department of Physics and
Cooperative Institute for Research in Environmental Sciences
University of Colorado
Boulder, Colorado 80309, USA

The last few years have seen exciting progress in geodesy, due, particularly, to the development and implementation of new observational techniques. As these techniques continue to mature and as the data continue to accumulate, geodesy will become increasingly important in helping to solve some of the outstanding problems in geophysics. Some of those problems are discussed here, with emphasis on the likely role geodesy will play in their solution. For this discussion, "geodesy" will include the observation and interpretation of: (1) the earth's gravity field, including both the time-dependent and the mean fields; and (2) the motion of points on the earth's surface at periods between a few hours and a few centuries. (This range of periods roughly corresponds to the useful range of geodetic instrumentation: shorter-period motion is detected more readily with seismic techniques, and the detection of longer-period motion must wait until there are observations over those long time spans.) Category (2) includes motion related to tectonic forces, the earth's rotation, earth tides, and certain long period normal modes of the earth.

Three important geophysical problems which geodesy can help address are:

Problem A: understanding mantle convection

Problem B: understanding the response of the crust and the lithosphere to forces associated with mantle convection

Problem C: understanding the core dynamo

Problem A

There are really two problems associated with understanding mantle convection. The immediate, kinematical problem is to determine the spatial pattern and time dependence of convective flow within the mantle. What are the horizontal and vertical scales of the flow? To what extent does the flow pattern correspond to the configuration of plates on the earth's surface? Is the flow steady or is it episodic, and over what time scales? To determine the spatial resolution, the most useful technique for the future is probably three dimensional seismic imaging. Gravity observations will be useful to help confirm the seismic results (gravity observations are sensitive to density anomalies associated with the flow), and they will continue to help suggest further details of the flow pattern in regions where there is insufficient seismic resolution. But, because gravity observations, no matter how complete, cannot uniquely determine the earth's internal density distribution, seismic data are inherently more useful. Any ambiguity in the interpretation of seismic results is due to insufficient data quality, not to any fundamental non-uniqueness of the technique.

Seismic techniques are less useful for determining the time dependence of the flow. Geodetic point positioning techniques, on the other hand, are potentially very important. Using these techniques, the geodesist can detect motion of the plates as it occurs in real time. There are, of course, possible ambiguities in the interpretation. For example, to what extent is the crustal motion really the surface expression of the underlying flow? And, how well can plate motion be separated from plate deformation? But, if these questions (both of them of interest in their own right) can be answered, the geodetic results can give information about the motion on two time scales. They can be used to detect episodic motion occurring during the time span of the observations: from years to decades. And, through comparison with plate motion models based on geological data, they can help indicate whether the motion has changed significantly over geological time.

The second and, in some sense, more fundamental problem related to mantle convection is the dynamical problem of modeling the flow. A complete numerical solution is, at present, a distant goal. The equations are strongly non-linear, and the relevant dynamical and material properties in the interior are poorly known. Geodetic observations can help constrain the relevant material properties, particularly those connected with the earth's anelasticity. Seismology gives information about the earth's anelasticity, but only at periods of less than an hour, far too short to be directly relevant to convective problems. Geoid anomalies and geodetic observations of the effects of post-glacial rebound have been used to learn about the earth's radially dependent, long-period viscosity structure. By using improved gravity results in the future, it may be possible to learn about lateral variations in viscosity and, conceivably, the dependence of viscosity on temperature. Furthermore, as earth rotation and earth tide observations improve, they should reveal more information about the earth's anelasticity at periods between the seismic and post-glacial rebound regimes. These intermediate-period results could help in identifying the mechanisms responsible for anelasticity, and this, in turn, could further help in understanding the long-period behavior of the earth.

Problem B

An important motivation for much of the ongoing work in observational geodesy is the problem of understanding the characteristics of crustal deformation in tectonically active areas. This includes learning more about earthquakes and the earthquake cycle, and studying the origin of geological features. These are among the most promising applications of the new geodetic techniques.

Problems related to the earthquake cycle can be addressed by observing the deformation as it occurs. The results can be used to constrain models of coseismic and aseismic slip along faults, to learn about the long period elastic and anelastic material properties in the vicinity of a fault, to detect the accumulation of interseismic strain which ultimately leads to faulting, and to search for possible earthquake precursors that might be useful for predicting earthquakes.

Observed crustal deformation can have important geological implications, as well, since it is the accumulated deformation that is ultimately responsible for most geological structures. Still, most structural features either are not presently evolving, or they are evolving too slowly to be readily detected in real time. To understand the origins of such features, it is valuable to learn more about the structure at depth. Again, three dimensional seismic imaging is apt to be particularly useful in this regard, but gravity observations can be of considerable use, both for resolving possible ambiguities in the seismic results and for learning about the mechanical strength of the region.

Problem C

As in the case of mantle convection, there are really two problems associated with understanding the core dynamo. The kinematic problem of understanding the flow pattern and the magnetic field in the core is a long way from being solved. Currently, surface magnetic observations help constrain results at the core-mantle boundary, but there are few constraints on the flow or magnetic field in the interior of the core. If, as is presently believed, the lateral seismic velocity variations in the core caused by convective flow are negligible, then neither seismic tomography nor gravity observations are likely to yield useful constraints on the flow.

The dynamical problem of modeling the flow is also far from being solved. One obstacle is that neither the relevant material properties of the core nor the relevant core-mantle coupling mechanisms are presently well understood. Here, geodesy can help. Certain rotational and tidal motions of the earth are affected by core-mantle coupling. Improved observations could lead to improved understanding of the coupling, although those motions are only sensitive to the coupling strength averaged over the entire boundary. Material properties inside the core could be constrained in the future through observations of internal fluid core modes and rotational and translational modes of the solid inner core. These modes all have long periods (a substantial fraction of a day and longer), and so are well suited to certain geodetic observational techniques. Most of these modes have yet to be modeled successfully for a realistic earth. However, their frequencies should be sensitive to properties of the core not easily accessible to seismology, and this makes them potentially valuable for the future.

GEOPHYSICAL GEODESY: THE STUDY OF THE SLOW DEFORMATIONS OF THE EARTH

Kurt Lambeck

Research School of Earth Sciences
Australian National University
Canberra ACT 2601, Australia

Geophysical observations provide the principal source of information on the interior structure of the Earth. They include the travel times, amplitudes and frequencies of seismic waves, the flux of heat through the surface, and magnetic-field parameters. Geodetic observations of the external gravity field and of the shape of the planet provide further constraints. Together with physical and chemical arguments, these observations form the basis for estimating the properties of the Earth's interior and for constructing models that describe the evolution of this and, largely by extrapolation, of other terrestrial planets. Geological observations, including geomorphology and geochemistry, are the key to understanding the history of the Earth's crust; of deformation events, of times of metamorphism and igneous activity and of horizontal and vertical movements. Palaeomagnetism provides the principal quantitative evidence for the large scale horizontal displacements and deformations, at rates of a few centimeters per year averaged over about 10^6 years. The examination of seismicity provides information on local and regional deformation of the crust on much shorter time scales. The central problem is the determination of the physical processes that led to these upheavals. What is the nature of the forces acting on the crust that have shaped it into its present state? What process is responsible for the large scale horizontal movements? What is the origin of the heat sources that produced the metamorphic and igneous events? The geodetic contributions to understanding these geological processes fall into two categories. The first is where observations provide a measure of the response of the Earth to known forces. In surface loading problems, where the applied force is known, observations of the deformation provide constraints on the rheological properties of the crust and mantle. In the second class of problems, the geodetic observations are used to provide constraints on the forces themselves. Many geophysical observations can be reconciled with radially symmetric Earth models, but the satellite results for the gravity field indicate that lateral structure is also important and these data point to a non-hydrostatic state within the Earth. These global gravity results reflect dynamical processes within the Earth and provide one further constraint on the Earth's present structure and, less directly, on its evolution.

The spectrum of the Earth's deformation is illustrated in Figure 1. At low frequencies and long wavelengths, the dominant deformations of the Earth are associated with mantle convection and its surface expression, plate tectonics. At the high frequency part of the spectrum, in both the dimensions of space and time, the dominant deformations are the earthquake displacement fields, the "instantaneous" expression of plate tectonics and mantle convection. The two extremes of the space-time spectrum are therefore closely related to each other, as well as to many of the intermediate wavelength-timescale deformations. The evidence for global tectonic motion is overwhelming, but models describing this motion are essentially kinematic ones and only rarely do they specify the dynamic mechanisms responsible for the surface motion and seldom do they specify the deformation beneath the crust or lithosphere. A central aim of modern geophysics and geology is to reach a quantitative understanding of the mechanisms involved and while considerable progress towards this goal has been made over recent years, it can hardly be said that the problem is fully understood. This inadequate state can be attributed to several factors of which two are particularly relevant: the limitations of our knowledge of the crustal and mantle rheology, and the limited understanding of the nature of the tectonic forces acting within the planet. Considerable geophysical research is directed, therefore, not only towards these global problems but also towards solving rather specific problems, including the refinement of seismic models of the Earth, the determination of the rheology of the crust and mantle, and the examination of tectonic processes at plate boundaries and within plate interiors.

With the broad definition of geodesy to include the study of crustal motion, the spatial and temporal variations in the gravity field and the tidal and rotational deformations of the Earth, the geodetic observations play an important role in the study of the structure and evolution of the Earth. Figure 2 summarizes specific geodetic techniques that may be applied to examine the deformations summarized in Figure 1. At the very long

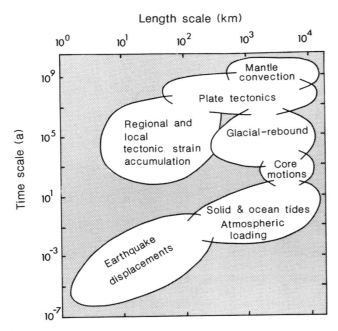

Fig. 1. The space-time spectrum of geodynamic processes leading to deformation of the Earth.

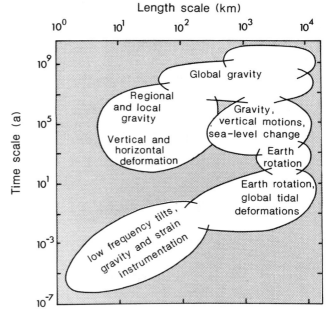

Fig. 2. The geodetic techniques relevant to measuring the deformations illustrated in Figure 1.

time scales, the observations of gravity and geoid anomalies provides information on the non-hydrostatic stress in the Earth and on the response of the Earth to this stress. For example, observations of gravity over seamounts or sedimentary basins provide a measure of the response of the lithosphere to loading on time scales of 10^6-10^7 years. On time scales of 10^3-10^4 years, observations of apparent changes in sea-level provide estimates of the response of the mantle to the extensive deglaciation that occurred between about 18000 and 6000 years ago. On shorter time scales, of days to decades, observations of the Earth's tides and rotation provide estimates of the global elastic and inelastic response of the planet. Locally, the geodetic observations provide estimates of crustal deformation that may otherwise go undetected because the seismic instrumentation response is insufficiently broad. Measurements of tilt and strain also elucidate the deformation process before and after the time of earthquakes. Even if there is not a geodetic remedy for every geophysical problem, there are many instances in which the observations play a supporting role. It must be recognised, however, that these observations provide information for only a very short segment of the Earth's history, of at most 100 years. This interval can sometimes be extended by using historical and palaeontological observations, as in the case of the Earth's rotation, or by examining geomorphological evidence, as in the case of variations in sea-level relative to present shorelines. This extrapolation is indeed necessary to bridge the time gap between the "instantaneous" geodetic measurements of deformation on the one hand and those geological and geophysical measurements of motion and deformation that represent average rates over geological time scales of 10^6 years or longer.

The new geodetic measurement techniques include the precise tracking of satellites, both artificial and the Moon, the long-baseline interferometric observations of radio sources, and the measurement of the Earth's gravity field using various sensors mounted on satellites. These new measurement procedures have not yet made a major impact on the geophysical discussion of the Earth's dynamic character, in part because these technologies have themselves been evolving rapidly and homogeneous data sets do not yet exist for even a few years. Yet the geodetic developments of the past two decades have now reached the point where one can see geophysical signatures arising out of the noise from only relatively short series of observations and one example of this is the VLBI observation of the Earth's short-period nutations. The promises that proponents of the new measurement methods have been making for many years, as discussed, for example, in the still highly relevant "Williamstown document" (Kaula 1970), are now being delivered.

As the accuracy of the space-geodetic measurements has increased, so has the complexity and cost of the equipment and it may well be asked whether all techniques that have been developed are necessary. Is it required to track artificial satellites with both lasers and electronic systems such as GPS? Is it necessary to develop and apply both laser ranging to the Moon and VLBI? The answer to these questions is yes. The GPS system does not, at present, preserve its inherently high precision when the station baselines exceed 100 km or so, and laser tracking is very complementary to GPS in crustal deformation studies on regional and continental scales. High precision laser tracking of close Earth satellites also provides the principal information for measuring the long wavelength part of the gravity field and on the time dependence of this field caused by tidal, meteorological and glacial rebound deformations. These observations are, however, less suitable for measuring the long-period components in the Earth's rotation because of the difficulty of modelling with high accuracy the secular and slow periodic perturbations in the satellite motion. Here VLBI methods are more appropriate in that the astrometric reference frame is well defined and in that these measurements have the potential for very high resolution observations. For the measurement of the global plate motions the VLBI methods appear to be at least as satisfactory as the satellite tracking methods, insofar as any conclusions can be drawn from only the short series of data that are now available. The lunar laser ranging methods would be the least satisfactory of the new methods for measuring the kinematics of the Earth, but it is the only method presently available for observing the orbital and spin motions of the Moon.

It would be hazardous to predict where our understanding of the deformations of the Earth will be when several decades of precise observations are available. New responses to known driving forces are likely to be discovered and as yet unknown mechanisms will be postulated. One reason why this prediction is hazardous is that the level of observations are now such that they are much contaminated by environmental factors, ranging from the meteorological excitation of the Earth's rotation to the role of rainfall and groundwater variations in crustal tilt measurements. What will be required in order to exploit fully the new geodetic measurements is a parallel observation program of the appropriate regional and global atmospheric-oceanic-hydrologic parameters. Another reason why the prediction is hazardous is that progress will undoubtedly occur in complementary areas of geophysics that require revision of today's concepts. Advances in seismic tomography will be very substantial once global and regional digital broad-band seismic networks are in full operation, and this will undoubtedly change today's dogmas on mantle convection and the interpretations of the gravity field will have to be adjusted accordingly. In some areas, the geodetic data has outstripped complementary geophysical data. For example, measurements of the flux of heat out of the Earth's interior are poorly distributed and improved gravity data will not contribute greatly to understanding the relations between these two geophysical quantities unless there is significant progress in the collection and interpretation of heat flow data. Even the surface topography is now less well known over many parts of the world's oceans than is gravity or geoid height, and future high resolution gravity field studies, using satellite-to-satellite or gradiometry techniques, could produce a similar state of affairs over large continental areas. Significant progress in understanding the Earth's gravity field will, therefore, come about only if knowledge of these other geophysical fields is also improved. One prediction that can be safely made is that the interpretation of future geodetic measurements of the Earth's deformation will require an increasingly deep understanding of geological and geophysical processes and that the measurements will become an increasingly integrated part of the Earth Sciences.

A few suggestions about anticipated developments over the next decade can be made with some assurance if only because the lead time of new satellite missions is almost this long. One area for ongoing geodetic research will be the Earth's gravity field. The solutions of the spatial variation in gravity, as derived from the analysis of perturbations in satellite orbits and from the analysis of satellite altimetry data, represent one of the major geophysical successes of the space program and these observations have played a very significant role in studies of mantle convection and lithospheric dynamics. Modest improvements in the knowledge of this field can still be expected, even if no new spacecraft are launched, through improved analysis techniques and through further precise laser tracking of some of the existing reflector-carrying satellites. A launch of additional satellites of the STARLETTE type, as has been mooted by the French space agency Centre National d'Etudes Spatiales, would also lead to useful improvements in the accuracy of the long and intermediate wavelengths in the gravity field. These developments would not, however, give significantly higher resolution of the gravity field over the continents than is now available and to achieve this a very major program will be required that is centred on technologies such as satellite-to-satellite tracking or satellite-gradiometry, as has been proposed by both the European Space Agency (ESA) and the U.S. National Aeronautic and Space Administration (NASA). A major value of such a mission is to provide much improved gravity information over active tectonic provinces at plate boundaries such as Tibet and the Himalayas or the Andean Cordillera. Altimetry data collected by the satellite GEOSAT and by the proposed joint ESA-NASA TOPEX mission will lead to an improved knowledge of the gravity field over the oceans but their major scientific contribution will be to the oceanographic studies of the time-dependence of the ocean surface and the associated forces and motions.

The study of the Earth's tides will remain an area of study where major progress can be made, both in the modelling of the ocean tides and in the estimation of the effective tide parameters from the analysis of perturbations in satellite motion. Altimetry data from the GEOSAT and TOPEX satellites will make major contributions to the former problem and additional STARLETTE-type satellites would contribute greatly to the latter problem. The principal contribution of such studies would be improved estimates of the rate of energy dissipation in the oceans and solid Earth and the estimation of the frequency dependence of the Love numbers due to both mantle anelasticity and fluid core effects.

Another important development will be the establishment of geodetic networks on the scale of tectonic plates for the measurement of crustal motions, probably with laser ranging and VLBI methods to establish regional frames and with GPS-type observations for the densification of the net that is required for understanding the strain cycles of the crust. The studies initiated in the eastern Mediterranean, centred on mainland Greece and the Aegean Islands, provides one example of such a combination of techniques that appears to be appropriate for the study of crustal deformation at complex plate boundaries. Measurement of time-dependent variations in height will be important in these programs and there will be a need for precise and long-term monitoring of sea-level as well as for geomorphological and geological studies of past motions. The continuous monitoring of motions of the major tectonic plates, with the methods of VLBI and satellite laser ranging to satellites, will provide an invaluable data base for analysing the plate motions to complement the geological estimates that are representative of average conditions over the intervals of 10^6 years. The importance of these "instantaneous" deformation measurements increases with the duration of the observation series and a minimum observation period would be the characteristic time interval between very large interplate earthquakes.

The Earth's variable rotation is another area of study where much progress can be anticipated as the records of the highly accurate observations increase in duration. This includes the nature of the excitation of the Chandler wobble but supplementary geophysical and meteorological information, such as atmospheric winds and surface pressure and the displacement fields of large earthquakes, will also have to be measured with improved accuracy and temporal resolution. Here the study of the Earth's rotation and crustal deformation is very complementary. Meteorological noise in the variations in length-of-day is now at least as great as the measurement noise and it represents a major limitation to the interpretation of the high frequency oscillations. Only with meteorological corrections will it be possible to examine questions such as the frequency dependence of Love numbers with any degree of confidence that the results are meaningful. Improved knowledge of the long period zonal ocean tides is also essential. New insight into the decade scale variations in both polar motion and length-of-day will be slower in coming, for the supportive information required for evaluating possible excitation mechanisms includes the time dependence of internal geomagnetic field parameters and the mapping of the topography of the core-mantle boundary by seismic methods. Here gravity field studies are complementary to the rotation studies.

These are only a few examples of matters that require further study if the newly emerging highly accurate observations of the Earth's deformation are to be understood. Many other examples could be given (Lambeck 1988). The really exciting work is only beginning.

References

Kaula, W.M. (ed.), The terrestrial environment: solid Earth and ocean physics, NASA Contract Rep., CR-1599, 1970.

Lambeck, K., Geophysical Geodesy: The Slow Deformations of the Earth, Oxford University Press, 1988.

SOME POSSIBLE ADDITIONAL ANSWERS

Petr Vaníček

Surveying Engineering, University of New Brunswick
Fredericton, N.B., Canada

Abstract. A few possible additional answers to questions posed in Moritz's paper, "Geodesy and geophysics in their interaction with mathematics and physics, and some open problems in geodesy," are suggested. These are presented within the context of future challenges for geodesy as perceived by this author.

Introduction

When the editors of this volume asked me to review Helmut Moritz's "crystal ball" paper I had taken it for granted that they wanted me to look at the problems addressed in that paper from a North American perpective. The opportunity of presenting the maverick New World alternative perception of outstanding geodetic problems side by side with Moritz's classically European point of view was too tempting to resist. While at it, I have come up with a few additional thoughts on the interrelations between geodesy on the one side, and mathematics and physics on the other. It is my hope that the reader will understand the contribution as a humble attempt to offer within this volume a broader spectrum of views rather than a criticism of Moritz's treatment.

Future of Geodesy

Let me begin with what are widely thought to be the main challenges for geodesy in the future on this continent. The typical North American approach to geodesy is very pragmatic. It is not automatically assumed that what is good for geodesy is good for society. Being forever forced to justify the position of our old science among the younger and rigorous sister geosciences, geodesists have to do so on the basis of usefulness to society.

It looks certain that the days of geodesy as a servant of mapping are numbered, regardless of what the future of mapping will be. The determination of positions to an accuracy needed for traditional mapping control is becoming a simple task. Due to the ever increasing role of various satellite positioning systems, specialized geodetic knowledge will soon no longer be needed for positioning for mapping control. Yet, positions are now determinable more quickly, cheaply, and accurately than before. This trend is certain to continue, viz my writing elsewhere in this volume. Positions are now a commodity sought out by an increasing number of customers among which solid earth geophysicists rank as the most prominent. The evolution of a customer pool will result in a significant change of the role for geodesy in society and a rise of completely new geodetic problems.

Increased accuracy of positions will force geodesists to take earth's temporal deformation effects on positions seriously almost everywhere on the surface of the earth. As it stands, only the most conspicuous deformation, i.e., large coseismic and ground settlement deformations, are deemed to affect positions severely enough to deserve any consideration in practical positioning. In the future, maintenance of up-to-date positions will require that a use be made of models for various deformational phenomena; not only body tide and sea tide loading but also of other phenomena such as global tectonics, sediment and volcanic loading, post-glacial rebound, and ground compaction. The challenge for geodesists will be in using these models and in cooperating with geophysicists in designing them.

Up until recently, geodesists had spent almost all of their time studying only 28% of the earth's surface: the land. Steadfastly refusing to acknowledge that even the positioning component of navigation has anything to do with geodesy, we had been very much "solid earth geodesists." This situation is now changing. With the exploration and exploitation of oceans, positioning and gravity field studies are being extended out into the sea. Geodesists, out of necessity, are becoming interested in problems of sea dynamics (effects on earth's rotation and on levelling datums), sea surface topography (effects on levelling datums), and the shape of the sea surface in general (indicative of earth's gravity field). These problems appear to be easier to tackle now with available tools such as satellite altimetry. It seems likely that the geodetic role in oceanography cannot but increase in the future making geodesy a truly global science. It is easy to see that a whole host of geodetic problems will arise when geodesists start venturing into the hydrosphere in earnest.

The other area of application of geodetic techniques will inevitably be in the planetary science. First steps in this field have been taken in the past 20 years and much more is to come. Without a doubt the evolution of planetary geodesy will bring about many, as yet unforeseen, scientific problems.

Finally, the ever increasing importance of various satellite-based geodetic systems is responsible for two more readily identifiable problems: the atmosphere, and satellite orbits. Ionospheric delay and the "wet component" of tropospheric delay are the major stumbling blocks in the domain of microwave ranging to and from satellites. To realize the full potential of this convenient and widely used measuring technique, atmospheric models will have to be significantly improved. Geodesists should think about spearheading the effort towards this goal. More about this point later.

Very accurate satellite orbit prediction, at the 1 cm level, is both desirable and achievable. To achieve this capability will require a development of better gravity field models, as stated by Moritz, as well as better models for solar radiation pressure on satellites. Better models for electromagnetic drag, earth's reflectivity (albedo) variations, and earth's deformation induced changes in the gravity field will have to be also developed. Satellite to satellite tracking and gradiometry should prove to be of considerable help in calibrating these models.

Geodesy and Mathematics

Speaking about the relation between geodesy and mathematics, the often used saying has it that nowadays a scientific discipline does well if the mathematics applied in that discipline is only fifty years old. Now, mathematics represents a very large body of knowledge that spans fields as diverse as formal logic on the one hand, and partial differential equations on the other, from the number theory, through mathematical statistics, to numerical analysis. The bulk of new knowledge in mathematics concerns the internal development of mathematics and is not of much use in applications. But even in the region of applicable mathematics, should we make an effort to use the catastrophe theory, or fuzzy sets just because they are "young" tools?

It seems that rather than measuring the status of geodesy by the "youthfulness" of its mathematical tools, it may be more useful to look at the breadth of mathematics that is being applied in our science. From this perspective, geodesy fares rather well. Apart from tools from the natural fields of geometry and partial differential equations treated extensively by Moritz, one sees more and more often tools being used that come from less natural fields of mathematics.

This volume was conceived to emulate the example set by Hilbert at the beginning of this century. It is thus fitting to cite here at least one example from a field to which Hilbert contributed substantially: functional analysis. Techniques from functional analysis are now being applied routinely in geodesy. Moreover, a synergism of functional analysis, geometry, and mathematical statistics is emerging as a powerful tool for some areas of geodesy. In this context, one may foresee a fertilization of mathematics by geodesy: normed or metric spaces with a Riemannian geometry are used in adjustment calculus yet they have not been much studied by mathematicians.

Classical mathematicians feel a pang of terror at the thought of working with singularities — as opposed to just removing them. It had taken a long time for Dirac δ-function to be accepted into the body of "respectable mathematics" and one is not sure if generalized matrix inverses have ever achieved that level of recognition. Yet, very useful results are being obtained, not only in geodesy and geophysics, by working with singularities. There are many problems where sets of particular solutions to singular inverses can be delineated and parametrized. A systematic theory of singular inverses, including the role of constraints and parametrization would be very handy in our sciences. Should geodesists/geophysicists lead the way?

One could mention also some fairly advanced techniques from numerical analysis, mathematical statistics, and other fields of mathematics used quite routinely in geodesy. But the space is short.

Geodesy and Physics

The uninitiated onlooker may wonder why geodesy, being basically the science of earth geometry, should have anything to do with physics. The close connection between geometry and gravity, treated so nicely in Moritz's paper, represents the standard answer to such an unspoken question. For the last 20 years or so, "physical geodesy" has been the synonym for the study of the only physical entity geodesy was interested in: the earth's gravity field.

Times are changing. Geodetic satellites and other geodetic systems based on diverse physical principles have come into existence and with them the role of physics in geodesy. Closer ties which have evolved between geodesy and geophysics, geodesy and oceanography, geodesy and space science, have underlined the necessity for geodesists to appreciate physics now more than they used to in the past. Nowadays, physics permeates geodesy almost to the same extent as it permeates geophysics.

If geodesy is successfully to come to terms with the tasks outlined in section 2, geodesists will have to study atmospheric physics as seriously as they study potential theory. Fluid mechanics will have to be accorded the same high degree of importance as celestial mechanics; electromagnetic theory will have to complement traditional geometrical optics. Basic knowledge of the mechanics of continuum and possibly thermodynamics will become indispensable. Further "physicalisation" of geodesy is clearly in the books for the future.

THE EARTH AS A PLANET

William M. Kaula

Department of Earth & Space Sciences
University of California, Los Angeles

Introduction

It is good to discuss now and then what directions our science should take. These discussions should include the rationale behind the research, particularly because the planning of research is usually done at the behest of a government agency whose goals may not be identical to those of the scientific community. Hence, since science is inherently international, an IUGG symposium "Quo Vadimus" is the appropriate setting for such discussion.

My answer to the question in Figure 1 is, rather intuitively, that the proper goal of the scientist is to infer the causes of the phenomena he studies: "...to see the forces behind the details" [Holmes, 1913]. The ultimate cause of everything may be a "big bang", but this is a rather unhelpful explanation for an obscure and complicated bit of the universe like the Earth. Hence, our definition of "cause" is conditioned by the background adopted, implicitly if not explicitly, for the phenomena studied. I discuss this question below.

It is also desirable, before accepting it as an exemplar, to examine Hilbert's setting-forth in 1900 of a list of twenty-three problems for his fellow mathematicians. Geophysics differs from mathematics, in that it does not work from axioms, but must take the world as it is. Geophysics also differs from basic physics or chemistry, in that most important phenomena cannot be isolated in a laboratory. Finally, it differs from other sciences that are necessarily environmental in orientation-- such as astrophysics and biology-- in that it must deal with one case (or very few) for a much greater portion of its important problems (as is also true for geology and planetary sciences).

Thus, philosophically, geophysics is quite different from mathematics. Indeed, it can be, and has been, argued that precise definition of the Hilbertian sort may inhibit, rather than help, progress at a fundamental level, by prematurely forcing our perceptions into a mold. To quote the introduction to a treatise on igneous and metamorphic petrology: "Precise definition generally will be avoided" [Turner & Verhoogen, 1960]. A leader in artificial intelligence elaborates this rationale: "I do not like definitions in something we do not understand. A clear definition put you in a box" [Minsky, 1986]. Definition of problems definitely does help guide the majority of researchers in their choice of what to work on. But, as is well known to any historian of the geosciences, there have been zig-zags in its course because of fashions in problem definition. We always want to keep an open mind for possibilities outside our current working frame.

While many of our most valuable insights come from recondite theoretical analyses, there do develop sub-communities of researchers within a science who refine internal criteria of merit-- sometimes largely esthetic-- thus causing their efforts diverge from the main themes as defined by the observational challenges. This problem arises in part because the techniques required by the theoreticians are greatly different from those of the data gatherers, so that it becomes more difficult and less entertaining to communicate. I can recall attending monthly sessions on the construction of the radar altimeter on the Pioneer Venus spacecraft. These meetings reminded me of the painting of the creation of Adam by Michelangelo: the radar engineers and myself touched fingertips on only one thing in common: the distance from the spacecraft to the surface of the planet. It takes appreciable conscious effort, often tedious, to keep experimenters and theoreticians in touch.

Clearly, most geophysics is not driven by esthetic criteria; it shares the main drive of the natural sciences of seeking the causes behind the phenomena. Hence, if we try to hypothesize a super-geophysicist to perform Hilbert's role, he would have to possess a combination of talents. Like Hilbert, he would try to reason rigorously from a comprehensive internally consistent set of axioms. But, like Einstein, he would wish these

Why? Why anything?

Drawing by Ed Fisher; © 1970
The New Yorker Magazine, Inc.

1. "Why do we do what we do?" should be examined occasionally.

"*Careful, now. I don't like the looks of this.*"

Drawing by Ed Fisher; © 1977
The New Yorker Magazine, Inc.

2. We always want to maintain an open mind to new approaches.

axioms and the deductions therefrom to be consistent with the phyiscal reality. Furthermore, like von Neuman, he would realize that certain facts about this universe are inherently unknowable because of either complexity or inaccessibility, so that aggregate, or "gestalt", parameters-- quite different from fundamental physical laws-- lead to more fruitful insights. Finally, like Hubble, our super-geophysicist would assert the need to get out and look. (The great majority of geophysicists are Hubblean, rather than Hilbertian or Neumanian; very few are Einsteinian).

The foregoing cautions notwithstanding, we all use something analogous to a set of axioms. Merely to decide what to do next, we need ongoing assumptions about the world in which our modellings or analyses or observations take place: we need a paradigm, to use currently popular terminology. Most commonly cited nowadays as an example of a paradigm in solid earth science is plate tectonics: an easy set of rules about the manner in which parts of the lithosphere move with respect to each other. Everyone sensible agrees that plate tectonics is, at best, a rough approximation in many places, and that it is quite removed from the underlying causitive physics. A much older paradigm, but enduring, is the uniformitarianism of Charles Lyell: the sedimentological record is the consequence of the geological cycle-- the processes of volcanism, plutonism, uplift, sinking, erosion, and sedimentation happening now.

A generalized uniformitarianism is applied, consciously or implicitly, in nearly all geophysics and related sciences of the natural environment: assumptions that all members of a class of phenomena are the consequences of the same underlying processes that vary-- temporally or spatially or both-- much more gradually than the occurrence of the phenomena. Table 1 enumerates examples of such uniformitarianism. An important variation in Table 1 is the nature of the uniformitarian background. The classical one of Lyell is largely observed directly, and rather firmly relatable to the phenomena explained. Some backgrounds have never been seen, or, at best, through a glass darkly, such as star formation. Others are things well seen, but not at all well explained, such as the plate tectonics already mentioned. Another example in this category is the quasi-stationary spiral structure (QSSS) of galactic dynamics, likened by Toomre [1977] to climbing a mountain by landing on a shelf partway up and working therefrom in both directions: "... the route downhill from B toward first principles A has since proved surprisingly treacherous,... On the other hand, the going uphill from B toward... C has... turned out to be delightful despite some crevasses. Of course, a nasty segment from C to D still lies ahead." As a general rule, the closer a paradigm is to phenomena, the more useful it is, but the greater the danger of error or oversight. Finally, we must not allow our paradigm to overwhelm the data: "It is the nature of an hypothesis... that it assimilates every thing

TABLE 1 Examples of Uniformitarianism

Class of Phenomena	Uniformitarian Background
Sedimentological Facies	Geological Cycle
Crustal Differentiation	Volcanism and Plutonism
Volcanism and Plutonism	Mantle Convection
Earthquakes	Plate Tectonics
Plate Tectonics	Mantle Convection
Terrestrial Planets	Planetary Accretion
Planetary Systems	Star Formation
Star Formation	Star-cloud Associations
Dynamic Evolution of Galaxies	Quasi-Stationary Spirals

to itself as proper nourishment; and... it generally grows the stronger by every thing you see, hear, read, or understand. This is of great use." [Sterne, 1760].

The solid earth is very much of one piece. Anything said about it assumes an evolutionary pattern to some degree. Table 2 gives a simple outline of the main stages of the Earth's evolution. On it are marked the average energy generation rates of each phase, to emphasize their appreciable differences in character. If we follow the dictum of seeking causes relentlessly, then we are driven far back in time. The major differentiations-- core from mantle, oceans and atmosphere from solid earth, crust from mantle, and continental crust from oceanic crust-- all took place, or started, quite early in the evolution of the Earth. If by "Earth as a Planet" we further mean explaining why the Earth is different from other planets, then we must examine a very different regime: the origin of the solar system, a relatively rapid process occuring more than 4.5 billion years ago.

The phases in Table 2 are arbitrary, defined for convenience of thinking. I have tried to remove some arbitrariness by indicating, by braces, phases that may have overlapped in time. Thus the inhomogeneity that led to Jupiter may have occurred rather early as part of the collapse of the protosolar nebula. The drastically varying degrees of retention of hydrogen and helium by the planets require that Jupiter (which retained 34 percent of its solar complement of H+He) formed before the other planets, and that the gaseous nebula was lost while the post-Jovian planets were accreting.

TABLE 2 Phases of Earth Evolution

Time Ago Ga		Energy Generation Watts/Ton	Order of Difficulty
	Collapse of Solar Cloud	60.	
	Formation of Jupiter		1st
4.57			
	Loss of Nebula		
	Planetesimal Growth		
4.56			
	Terrestrial Planet Accretion	0.0002	3rd
4.53			
	Hadean	0.000004	2nd
3.8			
	Arcadian		
2.0			
	Proterozoic		
0.6			
	Phanerozoic	0.00000007	
0.0			

Problem selection, in the Hilbertian spirit of this symposium, is a further arbitrariness. But we can say some inferences of Earth history are more difficult than others, and hence more appropriate to articulate as challenging problems. I give my estimate of the order of difficulty in Table 2. In the dynamics of planet formation, the great unkown is the formation of Jupiter, and the frequency of formation of such a companion to a solar-sized star. Once there is a proto-Jupiter-- necessarily while a massive gaseous nebula is still present-- there are sweeping resonance mechanisms to account for the widely spaced, seemingly over-stable array of post-Jovian planets. Inferring these mechanisms and the consequent pattern of planet formation involves considerable difficulties, but the problem is relatively well-constrained. As for the Earth's own evolution-- so different from Venus's-- the main conundrum is the initiation of continental crust

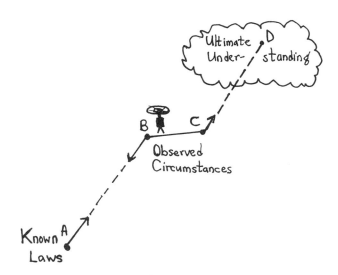

3. Paradigms that are well-observed but not well observed can be treacherous.

over a limited portion of its surface; why does the Earth not have a global continental crust, as Venus apparently does? Finally, there are differences between Venus and Earth that must be imputed to "original sin"-- the circumstances of formation-- most notably the differences in rotation rate (and the associated matter of the Moon) and retention of inert gases. While it is becoming the consensus that the final stages of terrestrial planet formation entailed great impacts, there are several aspects of this scenario that are ill-understood.

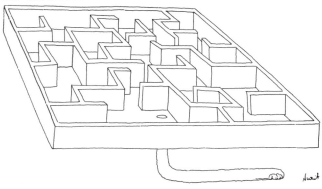

Drawing by Nurit; © 1975
The New Yorker Magazine, Inc.

4. Hilbertians should beware of leaky paradigms.

The three problems selected here all are instances of the question "How did this heterogeneity develop?". Furthermore, to understand these heterogeneities one must go a lot further than initial instabilities. Hence one looks to new ideas about chaos and other copings with highly non-linear systems, about which Dr. Keylis-Borok has spoken eloquently at this meeting. While techniques developed in one field can be useful in another, the transfer is never simple and direct, because of differences in the underlying physics. I am reminded of the words of another wise Russian, slightly modified: "All linear problems resemble one another; every nonlinear problem is nonlinear in its own way" [Tolstoy, 1875]. But we must grapple with these problems, if we are to understand the Earth as a planet; why it is so different from the other planets, including Venus, its closest sibling in size, composition, and distance from the Sun.

Origin of the Planetary System

The solar system-- more precisely, its planetary part-- is bothersomely unique: "Science can neither say nor do anything about a unique occurrence. It can only consider events that form a class" [Monod, 1972]. We do have a class of single main sequence G stars, of mass from 0.9 to 1.1 $M_\odot$ ($M_\odot$ is one solar mass), that, so far as can be observed, are quite close to the Sun in their properties. The nearest, 82 Eridani, is about six parsecs away. But, so far, there are firm indications of planetary companions for two K stars (0.5 to 0.7 $M_\odot$), inferred from their Doppler oscillations [Campbell et al, 1987]. Instrumentation beyond that on the Hubble space telescope, or the Space Infrared telescope facility (SIRTIF), are required for a significant advance. Hence, for the present, we have a problem suitable for computer experimentation.

Table 3 is a summary of the principal bodies of the solar system, in an arrangement intended to emphasize the great differences in their composition, which strongly constrains their mode of origin. It is clear that Jupiter is remarkable in two respects: its mass and its retention of hydrogen and helium. Models of Jupiter and Saturn indicate that they both have icy and rocky cores of about 15 Earth masses. In the fluid envelope around this core, Jupiter has retained about 34 percent of the H+He solar complement to the core, but Saturn has retained less than 10 percent; the other planets, less than one percent. It is thus seems ineluctable that Jupiter formed first among the planets; that this formation took place while an appreciable fraction of the gaseous nebula-- more than ten Jupiter masses-- was still present; and that both this proto-Jupiter and the gaseous nebula exercised significant gravitational effects on the material that came to constitute the other planets.

Hence our prime questions become: how unusual is a Jupiter-sized companion to a G-type main se-

TABLE 3 Planets by Type

Type & Name	Mass (Earth Masses)	Mean Density (g/cm^3)	Solar Distance (AU)	Average Orbit Inclination	Eccentricity
Fluid					
Jupiter	317.8	1.31	5.21	0.4°	0.045
Saturn	95.25	0.69	9.58	0.9°	0.048
Ice					
Neptune	17.21	1.66	30.20	0.7°	0.009
Uranus	14.60	1.19	19.14	1.0°	0.032
Ganymede	0.025	1.93	5.21	---	----
Titan	0.023	1.93	9.38	---	----
Callisto	0.018	1.81	5.21	---	----
Triton	0.0095	1.98	30.20	---	----
Pluto	0.0016	1.12	39.44	15.2°	0.242
Rock					
Earth	1.000	5.52	1.00	1.5°	0.030
Venus	0.815	5.25	0.72	1.6°	0.036
Mars	0.107	3.94	1.52	1.7°	0.08
Io	0.015	3.55	5.21	---	----
Moon	0.012	3.34	1.00	---	----
Europa	0.008	3.04	5.21	---	----
Iron					
Mercury	0.055	5.42	0.39	7.3°	0.18

quence single star? What are the processes by which such a companion can be formed? The answers to these questions undoubtedly control how often more extensive planetary systems form, the processes of their formation, and their resulting properties. Thus any model not taking into account an early proto-Jupiter and a gaseous nebula that was at least the complement of the planets in mass will be incomplete.

Current theoretical analyses and computer experiments on planetary formation are generally based on the model of a viscous accretion disk around the forming, or just-formed, Sun. The are several reasons for this approach: conceptual and analytical tractactability; the esthetic appeal of the mass-inward, angular-momentum-outward transfers in a viscous accretion disk [Lynden-Bell & Pringle, 1974]; the transferability of insights from those very different phenomena, planetary rings; limitations of computer capacity; and the possibility that any secondary heterogeneity developing before the Sun completed formation would be tidally disrupted. This work does contribute significantly to dynamical insight. But anyone engaged in this research would admit that it is very unlikely that the proto-planetary material and its gaseous complement aligned itself in perfect axisymmetry about the proto-Sun before God blew a whistle and proclaimed "O.K., let's make planets". This same community now says that an adequate mechanism for viscous transfer is lacking, and hence that gravitational interactions among asymmetric mass distributions are probably necessary.

It thus is still an open question when the heterogeneity essential to the formation of proto-Jupiter began, and how it began. Computer modeling [e.g., Boss, 1985] suggests that anomalously slow rotation of the proto-solar cloud fragment may be important. Physical plausibility suggests that heterogeneities have the best chance to get started in the early stages of collapse before opacity develops (at densities of about 10^{-14} g / cm^3) and temperatures rise, so that thermal effects squelch instabilities. But observation [Campbell et al, 1987] suggests that formation of a Jupiter is quite a distinct event from formation of a small binary companion: the same survey of sixteen nearby stars that strongly indicated two Jupiters found no indications of brown dwarfs in the range of 10 to 80 Jupiter masses. (" 'The dog did nothing in the nighttime.' 'That was the curious incident,' remarked Sherlock Holmes" [Doyle, 1894]).

We obviously would dearly love to have a big enough computer to trace the collapse of the proto-Sun in three dimensions, and to see what secondary inhomgeneities develop. Such computers are coming, sooner or later. Meanwhile, in the current intermediary phase of super-computer development, three-dimensional integrations of collapse are "...like cavalry charges: to be carefully planned, and saved for the right circumstances" [Whitehead, 1954]. A lot of work is needed not only in optimizing array-processor codes for such tedious essentials as radiative transfer, but also in problem definition: what hypotheses to test, what seed irregularities to plant. We cannot afford the unconscious hypothesis that nature developed instabilities the same as those developed by a computer from a homogeneous starting model through round-off error or whatever.

The Differences between Earth and Venus

It is evident from Table 3 that Venus is the planet by far the most similar to the Earth in its primary properties: mass, mean density, and distance from the Sun. Hence it is remarkable how different are the two planets in secondary properties: rotation rate, inert gas abundances, satellite existence, water content, surface temperature, topography, and magnetic field. Some of these are well-accepted as consequences of evolution. Thus Venus has a surface temperature $450°$ C higher than has the Earth (in contrast to the $17°$ C predicted from thermal equilibrium with the Sun) because its CO_2 is in the atmosphere rather than in the crust and ocean. This CO_2 maintains, and is maintained by, the greenhouse effect, whose onset could have been caused by a much smaller temperature difference between the planets. It also is concurred that if Venus had any water, it would all be sooner or later circulated to the top of this massive atmosphere, where it would be photodissociated, the hydrogen lost, and the oxygen combined with the surface rocks. The absence of a magnetic field is consistent with the energy source for the Earth's geodynamo being solidification of the inner core, since Venus is just smaller enough that the pressure at its center is less than that at the inner: outer core boundary of the Earth. The topography on Venus is unimodal: predominantly with rolling plains within one kilometer of the mean, with less than ten percent a few high plateaux, and no globally connected ridge system, as expected if plate tectonics prevailed as on Earth. This topography is consistent with solid Venus being completely enveloped in a thick crust, as discussed in the next section. The absence of a satellite around Venus can be explained as the consequence of tidal friction when the orbital period is shorter than the rotation period.

But two differences cannot be attributed to evolution: the rotation rates-- Venus's rate is - 1/243 times the Earth's-- and the inert gas retention-- Venus's $^{36+38}Ar$ is 80 times the Earth's. The difference in rotation rates clearly requires different impact histories in the formation phase. And it is quite implausible that the Earth would have less outgassing of primordial $^{36+38}Ar$, but more of radiogenic ^{40}Ar.

Such great differences between similar objects suggests that there can be large irregularities in their formation. Large irregularities suggest the chanciness associated with the statistics of small numbers. In the case of the formation of the terrestrial planets, the small number is the number of bodies preceding the final four. The scenario of this formation plausibly was a "Ten Little Indians" sequence: before there were four, there were five; before there were five, there were six; and so forth. In the particular instance of the difference between Venus and Earth, this scenario suggest that the Earth was hit by a sizeable impacter, while Venus wasn't: that the origin of the Moon and the loss of the primordial atmosphere implied by the drastic depletion of the Earth in inert gases (see Figure 5) both were consequences of a highly catastrophic event [Cameron, 1983].

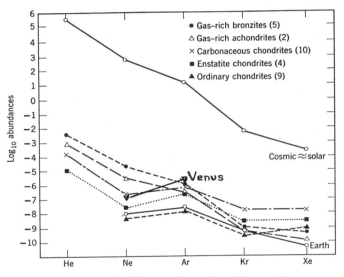

5. Inert gas abundances in meteorites and planets.

However, the great impact is somewhat of a hypothese faute de mieux for Both the inert gas depletion of the Earth and the origin of the Moon (not only its size, but its depletions in iron and volatiles): there are no plausible alternatives. The great impact is still unsatisfying to some because of its stochastic nature, or our inability to model it precisely, or the uncertainty of geochemical predictions such as differences between the Earth and the Moon in siderophiles. Hence on the one hand the great impact needs to be integrated into a scenario of the dynamical evolution of the inner solar system (which also accounts for phenomena such as the stunted growths of Mars and Mercury, the asteroid belt, and the slow rotation of Venus), while on the other, predictions of its consequences need to be worked out in detail. In particular, the closeness of the lunar $^{17}O/^{16}O$ and $^{18}O/^{16}O$ to those for ultramafic rocks in the Earth suggest that the impacter and the mantle were rather thoroughly mixed, so that compositional differences cannot be easily attributed to the Moon coming much more from the impacter. Hence they must be inferred from the impact process itself.

The logical compositional evolution of a terrestrial planet is a trend toward stable stratification. The aspects of the trend entailing differentiation of materials differing a few gm / cm^3 in density-- iron from rock, rock from volatiles-- apparently took place very early in planetary evolution (with the plausible exception of mantle-core separation in Mars). In some cases, most notably the Moon, nearly all separation of materials differing a few 0.1 gm / cm^3 in density-- calcaluminous silicates from ferromagnesian silicates-- also took place rather early. We suspect that crustal differentiation has ceased long since on Mercury, has dwindled to almost nothing on Mars, and may still be appreciable on Venus, from consideration of the effect of size on thermal evolution, and from their tectonics. But we know that crustal differentiation is still going on vigorously on Earth-- some 18 km^3/year. This 18 km^3/year is nearly all associated with the mid-ocean rises, a dragon 55,000 km long rising 2.8 km above the level of old (70 million years or more) ocean floor. No such dragon exists on Venus, as mentioned. The Earth is also extraordinarily effective at recycling this oceanic crust: dividing the 18 km^3/year into the estimated volume of oceanic crust gives a figure of only 110 million years; into the total volume of crust, oceanic plus continental, about 400 million years, less than 10 percent the age of the Earth. Neglecting that plate tectonics probably ran faster in the past because of higher energy generation, this means, crudely, that the continents are less than seven percent of the total of oceanic crust brought to the surface in the Earth's history. Hence it would not take much reduction in the efficiency of crustal recycling to lead to the entire surface of the planet being choked with continental crust within a billion years or so. This choking appears to have happened on Venus. Thus Venus is the more "normal" planet in its evolution toward the senility of complete stratification [Kaula, 1975].

The Earth is the strange planet in that it has, apparently throughout its history, recycled most of the oceanic crust. In Phanerozoic, this has clearly been by the process of subduction. It is difficult to imagine a significantly different process of recycling occurring in earlier eras. In Phanerozoic, we do have oceanic crust being subducted under oceanic crust, as well as continental, at places like the Marianas and Tonga Trenches. But it is all crust older than 100 Ma, and hence its initiation is explicable by models in which oceanic lithosphere and crust become unstable. If heat gradients are those expected to have existed in early Archean-- more than twice as great-- these models do not work.

The problem is how subduction got started, to lead to the configuration of a mix of continental crust-- low density (2.7 g/cm^3), much reworked material-- and oceanic crust-- high density (3.0 g/cm^3), more primitive material. It clearly got started longer ago than 3.8 Ga, the age of the stabilized rock associations, Greenschist belts, on southwest Greenland. It is difficult to imagine it starting earlier than the end of the accretion phase (defined as when the dominant energy source was infalls) more than 4.4 Ga ago.

We thus are left with a timespan of some 600 My in which to stabilize cratons large and buoyant enough to force subduction of oceanic lithosphere shoved against them. During most of this period, internal energy generation within the Earth (and Venus as well) was probably ten times as great as in Phanerozoic, while the average rate of mass infall was more like sixty times as great, as various quasi-stable dynamical wells were emptied. This rate-- which undoubtedly included impacts appreciably greater than that which created Mare Imbrium on the Moon-- still made accretion a minor energy contributor compared to internal sources during this phase. However, it had significant effects on the patterns of thermal convection and lithospheric tectonics, particularly because of its sporadicity. Hence it is easy to conjecture that even if there was a trend to stable stratification, the thermal (more than mechanical) pulses of great impact could have been sufficiently disrupting to allow plate tectonics to become established.

By why is solid Earth different from solid Venus? There are two external differences that are plausibly relevant: (1) surface temperature, apparently the residual of the one great impact blasting off the primordial atmosphere of the Earth; and (2) impact rates, because of differing distances from the quasi-stable reservoirs for impactors, the asteroid belt and the solar system beyond Saturn. The lower temperature on Earth acted directly to thicken the lithosphere and raise the basalt-eclogite transition, and indirectly to allow liquid water, which would reduce viscosities, reduce yield stresses, and enhance volcanism: all effects conducive to subduction of oceanic crust and formation of continental crust. The impact rates, according to most studies, would not have had greatly differing results for Venus and Earth. But this is dependent on assumptions about the provenance of the infalling bodies that may not be correct.

Hence to answer the question of how the lateral heterogeneities leading to stabilization of cratons evolved, allowing the remarkable Earth we have now, depends on the solution of a variety of subsidiary problems, which we can only enumerate briefly:
• the evolution of the planetesimal population that was the source of impacters in the Earth and Venus;
• the interaction of great impacts with mantle convection;
• the processes by which pieces of continental crust became paired with upper mantle low in large-ion-lithophiles and high in Mg: Fe ratio;
• the correct inferences from early Archean rock assemblages (whose very survival makes them anomalous) of the range of conditions in the 4.4 - 3.8 Ga period;
• the role of volatiles, especially water, in subduction and formation of continental crust;
• the differences in the physical conditions for crustal differentiation in early Archean from those in Phanerozoic.

The above list is undoubtedly incomplete and inaccurately stated. But these problems must be solved if we are to answer the question of why the Earth has such a remarkable lateral variation in crustal character, essential to many important aspects of the hydrosphere and biosphere.

References

Boss, A. P., Three dimensional calculations of the formation of the presolar nebula from a slowly rotating cloud, Icarus, 61, 3-9, 1985.

Cameron, A. G. W., Origin of the atmospheres of the terrestrial planets, Icarus, 56, 195-201, 1983.

Campbell, B., G. Walker, & S. Yang, A search for brown dwarf or planetary-mass companions to solar-type stars with high precision radial velocities. Bull. Amer. Astron. Soc., 19, 762, 1987.

Doyle, A. C., The Memoirs of Sherlock Holmes, Charles Newnes, London, 1894.

Holmes, O. W., Jr. "The Class of '61", Speech, Harvard, 1913.

Kaula, W. M., The seven ages of a planet, Icarus, 26, 1-15, 1975.

Lynden-Bell, D., and Pringle, J. E., The evolution of viscous discs and the origin of the nebula variables, Mon. Not. Roy. Astron. Soc., 168, 603-637, 1974.

Minsky, M. A., Radio interview, All Things Considered. Public Broadcasting Service, 1986.

Monod, J., Chance and Necessity, Collins, London, 1972.

Sterne, L., Tristram Shandy, D. & J. Diensley, London, 1760.

Tolstoy, L., Anna Karenina, pt. 1, Russian Messenger, Moscow, 1875.

Toomre, A. R., Theories of spiral structure, Ann. Rev. Astron. Astrophys., 15, 437-478, 1977.

Turner, F. J., & J. Verhoogen, Igneous and Metamorphic Petrology (2nd ed.), McGraw-Hill, New York, 1960.

Whitehead, A., Dialogues of Alfred North Whitehead, L. Price, ed., Little Brown, Boston, 1954.

SPACE PLASMA PHYSICS

Donald J. Williams

The Johns Hopkins University Applied Physics Laboratory, Laurel, Maryland

George L. Siscoe

Department of Atmospheric Sciences, University of California, Los Angeles

Born of the space age some 30 years ago, the field of space plasma physics has brought about a totally new and different perspective of extraterrestrial space. Far from being a passive void, space is now known to be filled with a teeming mixture of charged particles, neutral particles, and electromagnetic fields—a mixture known scientifically as a plasma. The plasma flowing from the Sun—the solar wind—extends to the farthest reaches of the solar system, where it interacts with its galactic counterparts. As it travels outward, the solar wind forms, at each planet in turn, an extended plasma environment—a magnetosphere—that is unique to that planet. Those turbulent magnetospheres, existing within and responding to the large-scale solar wind plasma, interact with planetary atmospheres, ionospheres, and surfaces to form a dramatic and important feature of each planetary environment.

When applied specifically to Earth, space plasma physics has come to be known as solar-terrestrial physics, which is defined as the study of the generation, flow, and dissipation of energy and the transfer of mass in the Sun-Earth system. This encompasses the study of the chain of intimately coupled regions extending from the Sun's surface to Earth—the solar photosphere, the solar corona, the solar wind, Earth's magnetosphere, Earth's ionosphere, and Earth's atmosphere.

By a synergistic fusion of elements from its root fields—geomagnetism, aeronomy, plasma physics, and astrophysics—solar-terrestrial physics has become a basic research field with well-defined scientific goals and a large and active constituency of researchers. Results of solar-terrestrial research continually are applied in those root fields, as well as in closely aligned research areas such as planetary magnetospheres and heliospheric physics.

The transformation of space over the past three decades into an important technological resource for all countries has made the results of solar-terrestrial research more important—not only do satellite systems carry out vital societal requirements in communication, remote sensing, and navigation, but colonies in space and transatmospheric flight loom as realistic future activities to be conducted in the harsh environment of space. The large variations in magnetospheric plasmas (which are stimulated by solar activity) are capable of harmfully affecting space and ground operations alike. Those variations—storms in space—have caused satellites to fail and to de-orbit early, ground systems to fail and to degrade ahead of schedule, and they pose a serious hazard for all manned activity in space. The need to predict "weather and climate" in space will become in the future as important as predicting atmospheric weather and climate.

Thus, as we look toward the future, we see that the major scientifc problem facing solar-terrestrial physics is to obtain a more complete understanding of the highly interactive solar-terrestrial system. We must understand its overall behavior well enough to develop a quantitative solar-terrestrial model that can predict the response of the system to varying solar input.

While ambitious, that is a realistic objective, largely due to the present level of maturity of solar-terrestrial research. In the first decade of the "space era" we saw the exploration of the distinct compartments of space, which led to important discoveries about the solar wind, the magnetosphere, and the ionosphere. In that first phase, each compartment formed a subdiscipline. The first theories were correspondingly parochial.

The late 1960s ushered in a second phase, as a movement to integrate the new science began when researchers realized how thoroughly the magnetic field in the solar wind regulates magnetospheric behavior. A new concept—solar-wind-magnetosphere coupling—grew to major importance. From a picture of territorial subfields separated by discipline fences, the conceptual level rose to give a view of linked domains—the "solar-terrestrial chain"—extending from the solar corona to the ionosphere. As in the traditional view of "solar-terrestrial relations," causality moved through the links unidirectionally from the Sun to Earth.

The onset of the third phase of solar-terrestrial research, during the 1970s, was marked by another rise in the conceptual level, when researchers recognized that the coupling between the magnetosphere and ionosphere was electrical and, therefore, two-way and interactive. The last link in the solar-terrestrial chain was replaced by a feedback loop, which ensured that the electrical currents leaving and entering the magnetosphere self-consistently matched the currents entering and leaving the ionosphere. Research at this new, more complex level needed computer modeling. In the ensuing decade (up to the present), numerical codes to model and simulate interactive magnetosphere-ionosphere coupling grew sophisticated.

The discipline has now evolved far beyond its early conceptual simplicity, its qualitative models, and its exploratory observations. Observations and theories have progressively revealed a more integrated and interactively coupled solar-wind-magnetosphere-ionosphere system. The movement has been away from regionalism and toward globalism, away from isolated processes and phenomena toward interactive coupling, and away from qualitative pictures toward quantitative models.

We have now reached the fourth phase; it began with the realization that, since the solar-wind-magnetosphere-ionosphere system is (almost completely) interactively coupled, it must be viewed as a unit. Because we have not been able to see the unit and watch it work—as we have with distant astrophysical objects—this realization dawned slowly through studying the separate parts and finding them connected and interdependent. In this fourth phase, the neutral atmosphere is seen to play an increasingly important role as the last interactive link of the solar-terrestrial chain. From what we know already, it is clear that a model having all of the main components interactively coupled is closer to reality than the old model, which had the components linked together serially. In the new model, causality operates in feedback loops to determine the system's integral response to solar variability.

Solar-terrestrial research must develop and test the solar-terrestrial equivalent of general circulation models so successfully employed in atmospheric research. To do so requires an integrated framework of theoretical modeling and experimental observation programs. It is here that international unions, associations, and committees play a vital role—in the coordination of existing and planned theory and experiment programs throughout the worldwide solar-terrestrial scientific community. For example, the ICSU Scientific Committee on Solar-Terrestrial Physics (COSTEP) has begun the organization of a Solar-Terrestrial Energy Program (STEP). National and multinational space programs will be placing observation platforms in key solar-terrestrial regions during the 1990s. Those observational programs encompass both space-based and ground-based facilities. Theory, modeling, and simulation studies are being planned to make immediate use of the data in the development of quantitative solar-terrestrial environment models. STEP will attempt to coordinate the many activities and encourage the development of efficient data distribution so that the concept of a general circulation model for space can be realized.

An important aspect in the realization of the quantitative models is the ability to obtain global information about the system. Much of our past information about the solar-wind-magnetosphere system has been obtained from *in situ* point observations. Those measurements have provided the foundation from which a good understanding of the detailed physical mechanisms operating within those systems has been developed. However, because of the physical scale of the regions under study, it is not practical to use *in situ* observations in a dense enough grid to provide a good global perspective of how the mechanisms work together to form the overall system.

New instrument developments now allow the magnetospheric charged-particle populations to be imaged on a global basis. Future observation programs will combine magnetospheric imaging with solar images, auroral images, and ionospheric remote sensing to yield comprehensive global measurements of the solar-terrestrial system. This would mark the first time that an astrophysical plasma system that has been well characterized by *in situ* measurements would also be characterized globally by remote sensing. Such a combination of observations would lead to a quantum leap in our understanding of solar-terrestrial system and would advance the development of a predictive solar-terrestrial model.

THE IMPORTANCE OF THE VARIABILITY OF THE SOLAR-TERRESTRIAL ENVIRONMENT

Y. Kamide
Kyoto Sangyo University
Kyoto 603, Japan

As Williams and Siscoe (this issue) point out, due to the combination of various data sets using various techniques with theories and modelling results, the last decade has seen a significant movement away from a concentration on regionalism or local structures and toward a focus on global patterns of solar wind-magnetosphere-ionosphere-atmosphere parameters, a shift from qualitative pictures toward quantitative models. The purpose of this report is to strengthen their conclusions by stressing (1) the importance of the nonlinearity of physical processes occurring in the space plasma system, and (2) the possible differences between average, empirical models of time-dependent phenomena and individual events.

1. Williams and Siscoe note that it is only recently that researchers have recognized that the coupling between the magnetosphere and the ionosphere is not "one-way" coupling but can be represented by a feedback loop. I would like to point out that the coupling is more than interactive and that the interactive processes are nonlinear in nature, thus generating "non-reversible" phenomena in the system. Magnetospheric substorms involving both suddenly initiating and rapidly growing auroral breakups are a typical example of such nonlinear processes. Furthermore, these sporadic phenomena must be explained self-consistently in terms of the generation, flow and dissipation of energy within the framework of the entire solar-terrestrial system, which WIlliams and Siscoe view as a unit. They emphasize the challenge of developing a quantitative model capable of predicting the response of the system to varying solar input. An inevitable complication lies, however, in the fact that temporal changes that occur in the real-life system are not merely the results of repetitions of quasi-steady states separated by Δt.

2. Because *in situ* techniques using point measurements have only a limited capability to provide a global picture with detailed physical causality, many statistical models have been developed. These models in fact provide an important foundation from which we can infer real observations as deviations from the average. However, the average picture sometimes -- or perhaps often -- misleads us about how the nonlinear processes occur within the large-scale system, particularly when the electro-dynamic state of the magnetosphere-ionosphere system is highly variable. As a simple example, imagine two disturbances with different spatial and temporal characteristics. The supposed average of these two disturbances may not exist in reality at all. In the magnetosphere-ionosphere system, these two may represent substorms and slow enhancements of the quiet state, which probably are, respectively, the manifestations of magnetospheric processes internally excited and those directly driven by the solar wind. We are just beginning to understand only the average state of our electrodynamic environment, in which many time-dependent and nonlinear processes are taking place. At present, we do not even know how sudden is sudden, although the onset of substorms is always described as a "sudden" increase in observed parameters.

Our ultimate goal is to understand in detail the whole solar-terrestrial coupled system or all the links within the system. Practically, our final goal may be to build a computer code that can predict, without relying on any empirical parameters, what will happen in space near the Earth. That is, we would like to be able simply to input the present condition of the solar wind, and then the computer's output would tell us how, when, where and what magnitude of substorms will take place, even how and where they will develop. Most importantly, however, before we start to write such a code, we must either know the physics or the equations governing the solar-terrestrial system or we must have the empirical relationships among a number of quantities that are crucial to the system.

The thirty-year history of the space age has taught us that, in the future, we shall encounter intriguing new problems in space physics -- problems of which we are at present entirely unaware. However, no matter what fascinating puzzles await us, we may still expect to arrive at the end of the next decade with a greater comprehension of both the large-scale and small-scale physics that govern the solar-terrestrial system.

WHERE ARE WE GOING IN THE STUDY OF SHORT-PERIOD CLIMATE FLUCTUATIONS?

Jerome Namias
Scripps Institution of Oceanography
University of California San Diego

The desire to predict short-period fluctuations of the order of a month, a season, or several years has probably been around for thousands of years. Such capability would enable man to plant and harvest efficiently, to plan water resources, fuel allocation and in general, prepare for all sorts of eventualities--both good and bad. The economic benefits to accrue from perfect forecasts are staggeringly great. But in spite of these facts, our ability to predict the nature of weather and climate for even the next month or season is only marginally successful. For still longer periods, it has not been demonstrated that any skill exists. These sobering facts suggest that here is one of the topics in crying need of better solution--and certainly is a subject requiring penetrating research involving physical understanding.

I stress that the route to go is via physical understanding, because over the past several decades and with accelerating pace, some have tried to short circuit understanding with the help of a seemingly infinite number of statistical techniques--each apparently designed to find answers through massaging of millions of data, often without regard to physical hypotheses or even rough concepts involving physical reasoning. One of the most simplistic of these shortcuts is the so-called analogue method, whereby one searches many meteorological fields for patterns like those of the present month or season, extracts the observed charts for that season, and then copies off the subsequent period's weather. Yet tests on this method carried out for many years, have yielded results of little skill relative to chance or simple persistence. By now, it should be clear that we are dealing here with a physical problem that will only yield to analysis in which the researcher employs the proper physical ideas for his edifice. As in experimental physics, any theory generated in this way can be empirically tuned with real atmospheric data. In the paragraphs to follow, we shall enumerate some of the problems to be solved. It is the hope that new generations will provide a better theoretical footing for climate prediction by using new sensing devices and observations, together with faster and faster computers and up-to-date technology. Experience in the short-range forecasting arena has shown that success is more likely if theoreticians work closely with empiricists. Geophysics is another field where such collaboration has been most fruitful.

We shall describe some of the pressing problems dealing with the subject of short-period climate variations and their prediction:

I. How good are present long-range simulations and forecasts and what methods are employed?

This topic has received a great deal of attention over the past couple of decades. Dedicated scientists have indeed gone far to simulate the mean climate with its wind and weather patterns for each of the four seasons and for individual months. Nevertheless, there are still large gaps in our understanding because of errors in certain regions and even in some of the statistics involving prevailing planetary wind systems. Some of nature's physical processes are still not sufficiently well understood to formulate mathematically, even though a good beginning has been made. We bring up this problem because it is generally acknowledged that before variations of the general circulation among different years can be predicted, we must be able to simulate the normal state accurately. Fig.1 shows an example of the current state of the art: a well-known simulation of global sea level pressure, using the up-to-date understanding of the physics of the atmosphere's general circulation, and the most sophisticated computers. This is a good estimate of the principal centers of action in the atmosphere, particularly in macroscale. However, careful scrutiny of this estimate shows that the magnitudes are not especially good in many places, particularly in the eyes of practitioners. For instance, the intensity and structure of the Aleutian Low, which is both an important weather generator as well as a weather indicator, is poor. To make better predictions and provide better physical understanding, errors such as these must be reduced.

Even if the simulation of the normal state were perfect, this does not insure that the departures from the normal for a month, season or year would be easily achievable; there would still be a big gap, and the ability to predict anomalous states would be lacking. This deficiency is the primary reason why empirical methods are usually employed in making long-range predictions.

An example of the current state of the art of monthly prediction is represented in Fig. 2, the U.S. National Weather Service's prediction for temperature anomalies over the contiguous United States for Spring 1987. These are

NORMAL SEA LEVEL PRESSURE

SYUKURO MANABE AND DOUGLAS G. HAHN

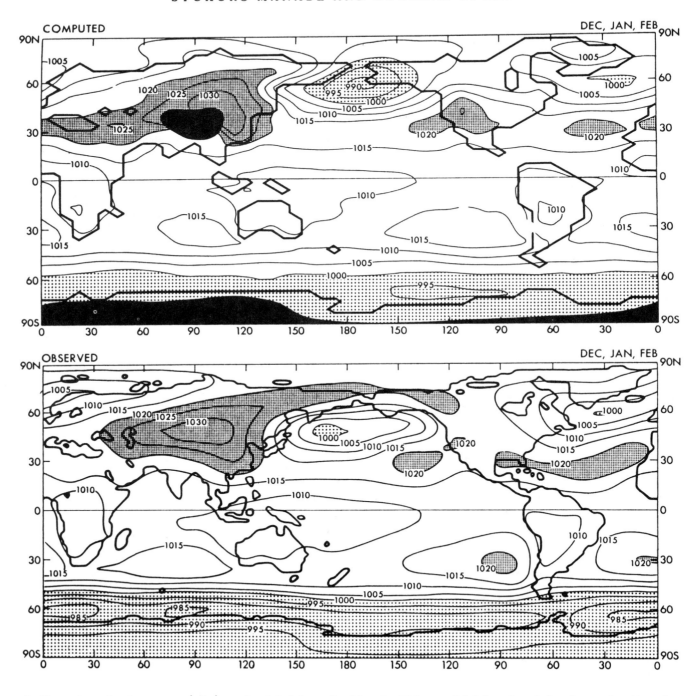

1. Normal sea level pressure (mbs) as simulated from the Manabe-Hahn model (above) and the observed pattern for the winter months (December, January and February).

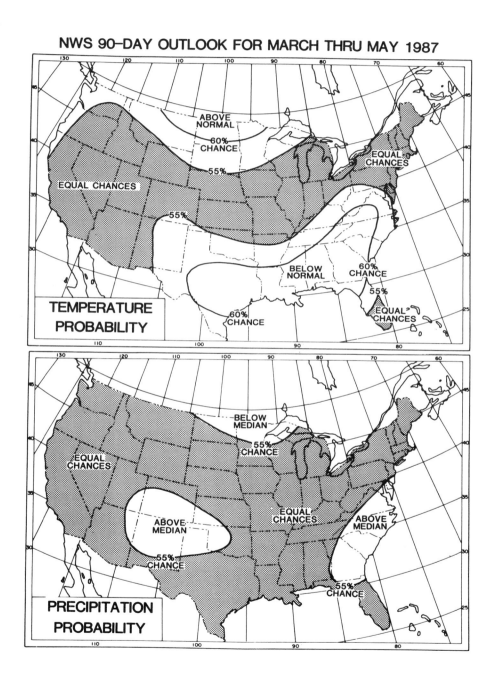

2. Predicted temperature anomalies for spring (March, April and May) expressed as probabilities, made by the United States National Weather Service's Long Range Prediction Group.

presently expressed in probabilistic terms--the probabilities being determined largely from past performance of forecasters. Judging from the values on this chart, skills have been only marginally successful. There is some question as to the desirability of using probabilities, rather than categorical forecasts and past performance. Also to be considered are the peculiarities associated with the case at hand.

Even for verification, entirely satisfactory solutions have not yet been found.

Another example of the shortcomings of long-range forecasting is supplied by Fig. 3, which shows the U.S. National Weather Service forecast for Europe (compared with the observed) for January 1987, a month characterized by extremely cold weather over all of Europe, extending to the British Isles. These persistently recurrent cold waves were associated with a phenomena called "blocking," in which the normally present westerly winds vanish and are replaced by an east-to-west drift that brought Siberian cold air masses progressively farther west.

Although blocking has recently received much attention in the scientific literature, our understanding is poor. Thus, as demonstrated by Fig. 3, its impact on European weather is often not foreseen. In this particular case, Western Europe was predicted to be warm (above normal), when in reality, it was extremely cold, with long-term records having been broken in many places.

Lest the reader get too negative an impression, I have chosen a few relatively good predictions--these on a seasonal time scale. These hold out hope for improvement with further research. Fig. 4 shows the result of one of a promising series of forecasts being produced by a sophisticated computer model. This is part of an experiment called "DERF" (Dynamic Extended Range Forecasting), being explored by a group of scientists at the Climate Analysis Center in NOAA, and other university meteorologists, including Dr. John Roads of the Scripps Institution of Oceanography Climate Research Group. He supplied me with Fig. 4, which shows remarkable success in predicting the average 700 mb height and anomaly patterns for the 30-day period from mid-December 1985-86 to January 1986. This was done using initial data from December 15, 1985, iterating forward, and averaging the day-to-day predictions for the entire 30-day period. Both the patterns and the anomaly predictions are strikingly good and capture the main features. It is possible to objectively specify what temperature departures are likely with this picture--shown in Fig. 5. Again, this would have been a valuable forecast. Although it is too early to reach firm conclusions, it appears that this new DERF procedure is the wave of the future.

An example of a good seasonal forecast is shown in Fig. 6, giving predicted and observed temperature and precipitation patterns for Fall 1986. This forecast was made by the U.S. National Weather Service, using more conventional tools: statistical probabilities from regression equations, long-period trends, analogues, air-sea interactions, contingencies and other factors. Of course, the weights assigned to these factors vary from case to case. Important in this case were some air-sea interactions in the North Pacific, south of the Aleutians, where cold water in summer frequently leads to higher than normal pressure in the subsequent fall, and the influence of this alteration of pressure spreads downstream to North America through well-known teleconnections. Fig. 7 shows the forecast and observed precipitation patterns for Winter 1982-83 and Spring 1983, a six-month period when strong El Niño conditions occurred. In these cases, the circumpolar westerlies were displaced south of their normal position, leading to a southerly displaced storm track over many areas. Such a track was foreseen this winter and spring, leading to good estimates of heavy precipitation over much of the contiguous United States.

It must be stressed that comparable ability to predict El Niños and their accompanying weather is not routinely achieved. Unfortunately, El Niños vary from case to case, especially their impact on the regional circulation and weather. Such variations are demonstrated in Fig. 8, which shows the precipitation patterns for ten recent El Niño winters.

Mentioned previously were the methods by which it is possible to specify temperature anomaly patterns from mid-tropospheric heights. These are stepwise multiple regression techniques called "screening," whereby predictors are determined empirically from a field (in this case, 700 mb heights over a large domain) and then translated with the developed equations into contemporary temperature anomalies for a close grid of cities. For the United States, the domain for selecting the predictors usually extends over thousands of miles. Because of the long wave "teleconnections" in the atmospheric flow, the temperature at Chicago, for example, depends on the circulation as far away as Alaska, as well as upon more local points. Unfortunately, mathematical-physical methods are not yet able to do as well as these statistical methods. Perhaps this is one of the primary problems facing deterministic long-range forecasting by physical means. A typical example of a temperature specification, for Spring 1977, is shown in Fig. 9. Obviously, if one could predict the upper level flow pattern with accuracy, a very good temperature forecast could be made for a month or season in advance.

Another statistical technique frequently employed in long range forecasting involves teleconnections--long distance interactions (cross-correlations) between upper level circulations involving the mid-tropospheric height values. Hundreds of these teleconnection charts, such as displayed in Fig. 10, have been worked up for each season and for points abut 10 degrees of longitude and 5 degrees of latitude apart. As seen from the magnitude of the cross-correlations, these teleconnections are highly significant. Their general existence has been known for at least a century, but it is only with the advent of high speed computers that they can be worked up as in Fig. 10, or indeed, in many other ways thought helpful in making predictions. Here again, dynamicists have a challenge to duplicate, and perhaps supplant, these charts with physically based teleconnections.

II. What are the influences of abnormal boundary conditions on the atmosphere and its general circulation?

This problem has come into focus in the past couple of decades following several empirical studies indicating important influences of ice and snow cover, of sea surface temperature anomalies, and of variations in soil moisture on behavior of the overlying and remote atmosphere. Perhaps most popular of these is the El Niño, wherein a massive pool of abnormally warm water straddling the equator

JANUARY 1987 TEMPERATURE/DN

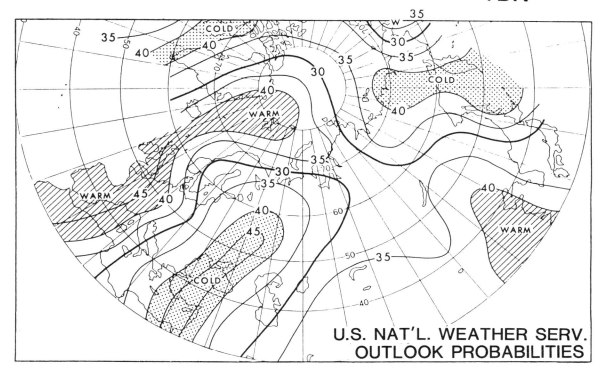

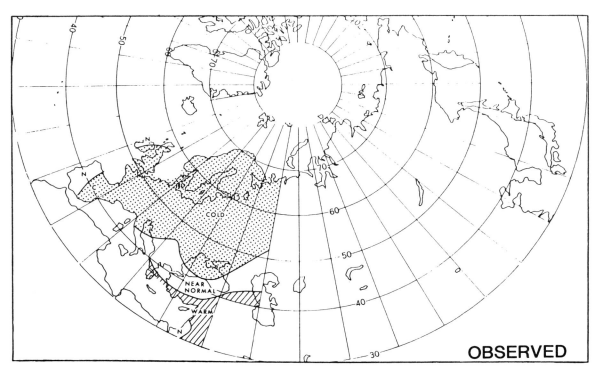

3. United States National Weather Service temperature forecast for Europe, expressed in probabilities, for January 1987 (above) and the corresponding observed temperature anomalies (below).

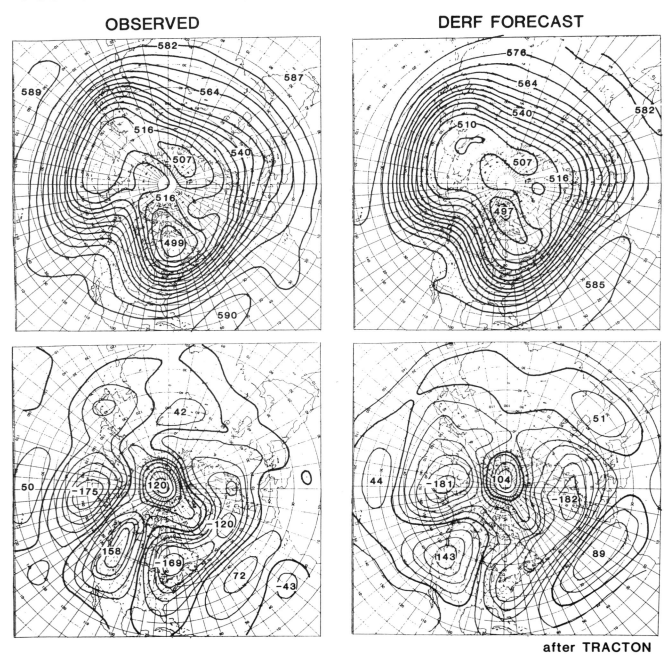

4. Prediction of the 700 mb pattern (in mbs) for the period mid-December to mid-January 1986 (above), made by DERF (Dynamic Extended Range Forecasting), together with the observed (below). (See text.)

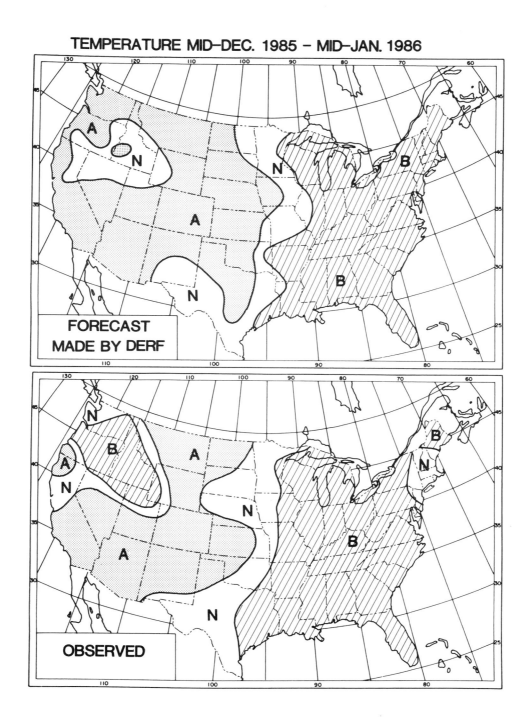

5. Temperature anomalies specified from the DERF prediction shown in Fig. 8, together with the observed anomalies.

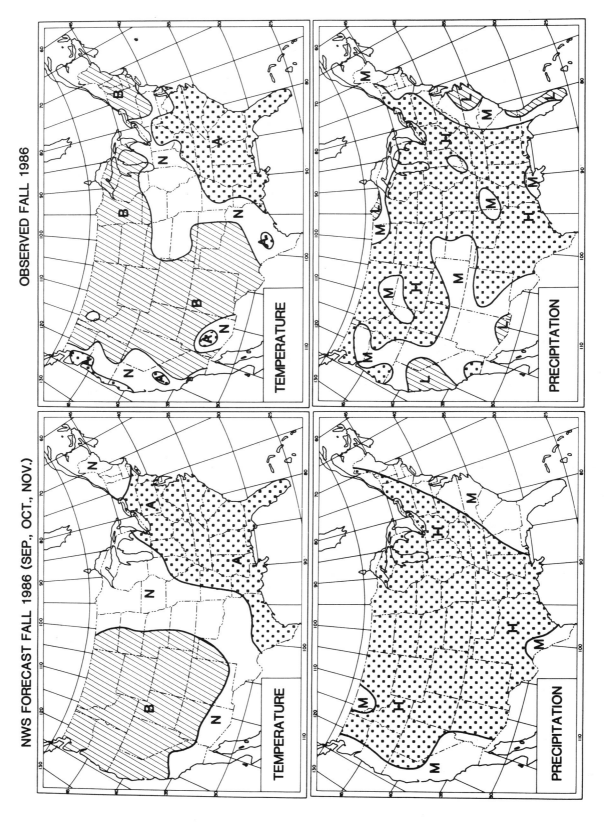

6. Prediction of temperature and precipitation anomalies by the United States National Weather Service for Fall (October, November and December) 1986, together with the observed patterns.

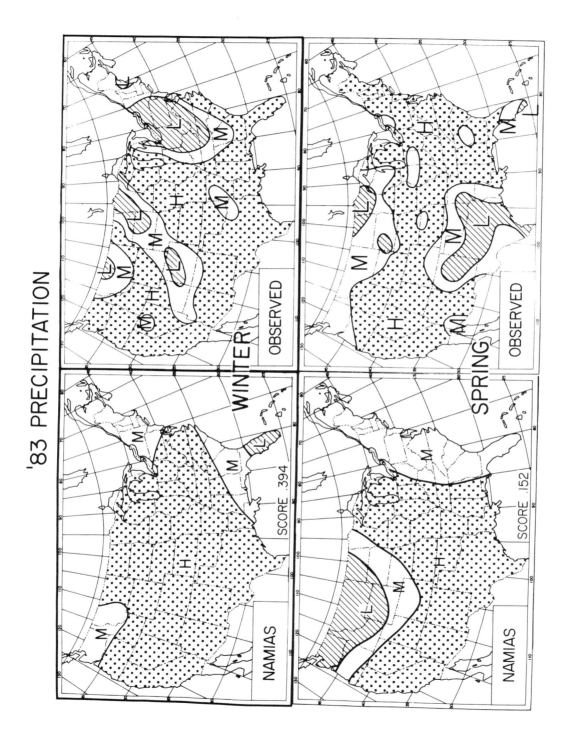

7. Precipitation forecast, expressed in terciles, Light, Moderate and Heavy, for the contiguous United States for Winter 1983 (an El Niño year), together with the observed pattern. Made at Scripps Institution of Oceanography, Climate Research Group.

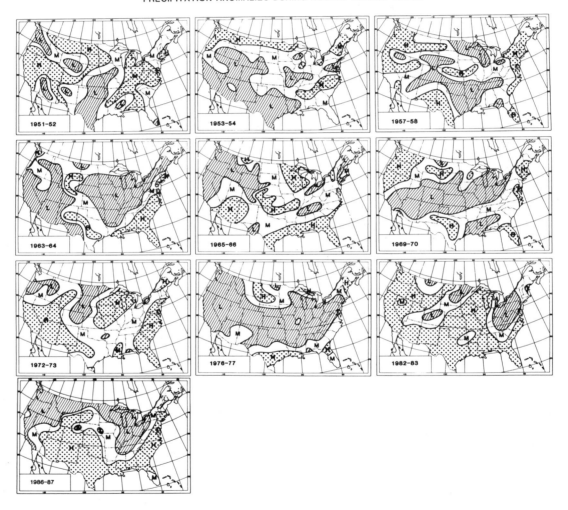

8. Precipitation anomalies for ten recent El Niño years, expressed in terciles: Light, Moderate and Heavy.

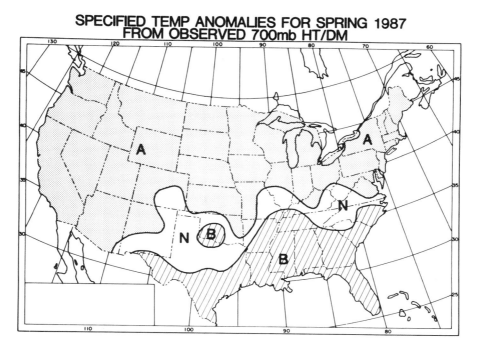

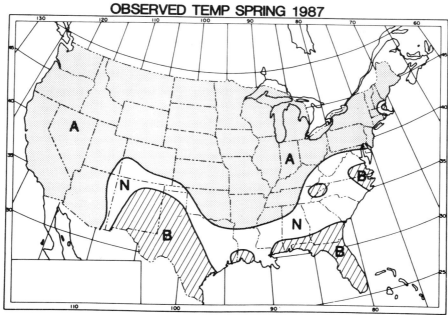

9. Specified (above, from the observed 700 mb anomalies) and observed temperature anomalies (below) for Spring (March, April and May) 1987.

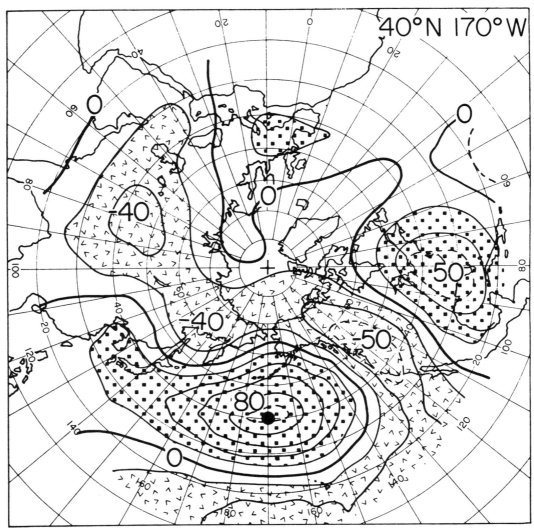

WINTER TELECONNECTIONS OF 700 DM

10. Winter teleconnections (cross correlations) from a point in the Central North Pacific (40°N, 170°W) to all other points over most of the Northern Hemisphere. Constructed from about 30 winters of data.

influences atmospheric convection in equatorial regions, and this in turn affects the planetary wind systems, even in temperate latitudes. Evidence suggests that the warm water of El Niño is, in fact, generated by abnormal wind systems in the tropics, where there is strong interaction between atmosphere and ocean. Tropical Pacific water temperatures are observed to have anomalies lasting several months, sometimes as long as two years. If this aberration could be predicted, climatologists would be able to predict some weather and climate characteristics in many places of the globe. Unfortunately, this *desideratum* is not at hand. To complicate matters, the phenomena of El Niño has different influences in different places and from one case to another. We are able to describe and measure many of the phenomena associated with El Niño, but prediction is another matter. This is another problem awaiting solution, needing the guidance of insightful dynamicists.

Influences of abnormal ice and snow cover also play important roles in short- and long-period climate variations; here again, many empirical studies and some numerical modelling efforts are encouraging. However, at present, direct use in forecasting is difficult.

Variations in soil moisture is another only partially solved problem. It is known that after a wet spring in the plains regions (especially in the United States), the increased radiation with the coming of summer is more likely to be utilized in evaporating moisture from the soil than in generating surface heating. The net result is that summer temperature is influenced; wet springs tend to be followed by cool, showery summers, and vice versa for dry springs. Numerical modelling sensitivity studies have begun on this topic, and results are encouraging for ultimately using the concept objectively in long range forecasting. In some quarters, it is believed that these boundary layer influences and their understanding comprise the most important problems to be solved for practical forecasting.

III. Abnormal weather and climate spells lasting a few years or more

The problems associated with long-period spells of years of abnormally similar weather and climate pose some of the most intriguing and economically important of all long-range forecast considerations. The difficulty in explaining these regimes may account for the lack of progress in this area; in fact, it is questionable whether any progress has been made. We refer here to phenomena such as the Dust Bowl that occurred in the 1930s, when recurrent drought dominated the great plains of the United States. Other parts of the world have been impacted by long-period aberrations from normal; the Sahel drought, for example.

The memory-producing devices, within or outside the atmosphere, that could account for these spells certainly have not been identified. This situation has led to hypotheses involving man-made (anthropogenic) causes, such as decertification in the case of the Sahel, or certain chemical interactions, such as the increasing CO_2 and the use of fluorocarbons. While studies of this nature must proceed, it is possible that these spells are, in fact, expressions of the same phenomena discussed earlier in connection with those lasting a month or season. Numerous influences might be invoked to explain these changes, even though they have not yet been proved or identified on such long time scales. Recently, efforts have been devoted to proposing long cycles for the El Niño that could conceivably rule weather patterns for spells of several years. Like other hypotheses, these are not yet adaptable to practical long-range forecasting.

For centuries, efforts have been made to show that these long periods swings are associated with the solar cycle, particularly, the 11-year and 22-year (Hale) sunspot cycles. The time scales are of the right order to raise this possibility. However, the consensus in the meteorological community is that solar-weather students have not proved their point, either empirically or theoretically. In spite of this, some scientists still adhere to these ideas. The recent improved knowledge of variations in the solar constant and its spectral variations with the help of satellites, and better data coverage of the globe should spark more research along these lines. Most solar-weather work is tantalizing but hardly conclusive. It is possible that numerical simulations will help solve this problem, even though vast quantities of data and the necessary iterative computations extending over many years look foreboding. These problems are so vital for mankind that they deserve the attention of the best scientific minds, as well as more open minds in the meteorological community. Perhaps some novel ideas will be found by which iterations need not be done day by day, but with short cuts using as yet undeveloped macro-turbulence ideas employing statistical ensembles.

IV. How best can man interact with machine to produce better forecasts?

The topic may seem out of place in this short summary on short-term variation in climates and their forecasting, but in the mind of the present author, it is extremely important if forecasts are to achieve a high level of success. The truth of the matter is that at present, long-range forecasting (and to a lesser extent, short-range forecasting) is as much an art as a science. This state of affairs exists in spite of the extensive research and apparently objective techniques developed. Because so many factors have to be taken into account in making forecasts, weighting becomes paramount. A forecast tool that may successfully be applied in one situation may be of little value in another. The decisions of selection of appropriate indications require much experience and practice with many types of situations. The aim of scientific research should be to develop more objective systems.

Of late, the subject of artificial intelligence has attracted many people, scientists among them, and some of these adherents feel that it may be applicable to long-range forecasting. This concept is hard for "old timers" to swallow because of the peculiar nature of the subject, which has to synthesize large quantities of material and deal with the interaction of many time and space scales. However, it is always possible that the development of faster and more sophisticated computers will rectify these difficulties.

The forecast problem involves not only our unsatisfactory state of knowledge, but also the question of inherent predictability of the atmosphere. Predictability is important on all time scales, but vital on the long-range scale. In the past decade, many sophisticated studies of predictability have been pursued. Some scientists have concluded that, at times, the atmosphere does not know what it is going to do next, and that its evolution at long-range could go into a number of stable modes -- not only one. In other words, an inherent indeterminism may be involved. Most practitioners are understandably opposed to these ideas and prefer to take a more optimistic approach. Nevertheless, the subject of predictability will probably be around for years to come.

LEGENDS TO FIGURES

1. Normal sea level pressure (decameters) as simulated from the Manabe-Hahn model (above) and the observed pattern for the winter months (December, January and February).
2. Predicted temperature anomalies for spring (March, April and May) expressed as probabilities, made by the United States National Weather Service's Long-Range Prediction Group.
3. United States National Weather Service temperature forecast for Europe, expressed in probabilities, for January 1987 (above) and the corresponding observed temperature anomalies (below).
4. Prediction of the 700 mb pattern (in mbs) for the period mid-December to mid-January 1986 (above), made by DERF (Dynamic Extended Range Forecasting), together with the observed (below). (See text.)
5. Temperature anomalies specified from the DERF prediction shown in Fig. 8, together with the observed anomalies.
6. Prediction of temperature and precipitation anomalies by the United States National Weather Service for Fall (October, November and December) 1986, together with the observed patterns.
7. Precipitation forecast, expressed in terciles, Light, Moderate and Heavy, for the contiguous United States for Winter 1983 (an El Niño year), together with the observed pattern. Made at Scripps Institution of Oceanography, Climate Research Group.
8. Precipitation anomalies for ten recent El Niño years, expressed in terciles: Light, Moderate and Heavy.
9. Specified (above, from the observed 700 mb anomalies) and observed temperature anomalies (below) for Spring (March, April and May) 1987.
10. Winter teleconnections (cross correlations) from a point in the Central North Pacific (40°N, 170°W) to all other points over most of the Northern Hemisphere. Constructed from about 30 winters of data.

FURTHER READING

Namias, J., 1953: Thirty-day forecasting: a review of a ten-year experiment. Am. Met. Soc., 2, No. 6, Meteorological Monograph, 83 pp.

____, 1968: Long-range weather forecasting -- history, current status and outlook. Bulletin of the Am. Met Soc., 49, 438-470.

Ryan, P.R. (Ed.), 1984: El Niño. Oceanus, 27, 84 pp.

Tracton, M.S. and Kistler, R., 1988: Activities and dynamic extended range forecasting (DERF) at the National Meteorological Center. Presented at the Numerical Weather Prediction Conference, AMS. In press.

Tribbia, J.J. and R.A. Anthes, 1987: Scientific basis of modern weather prediction. Science, 237, 493-499.

Walsh, J.E., 1983: Sea ice, snow cover and soil moisture. Proceedings of the WMO-CAS/JSC Expert Study Meeting on Long-range Forecasting (Princeton 1982). World Meteorological Organization, Long-range Forecasting Series #1, 84-96.

GEOPHYSICAL FLUID DYNAMICS AND RELATED TOPICS

Raymond Hide

Geophysical Fluid Dynamics Laboratory, Meteorological Office
(Met O 21), Bracknell, Berkshire, RG12 2SZ, England, U.K.

Extended summary of an invited talk at the Vening Meinesz Centenary Symposium "Quo Vadimus" (U1) held on 13 and 14 August 1987 at the 19th General Assembly of the IUGG, Vancouver, Canada.

Introduction

In preparing the programme of this symposium, the organizers must have asked many geophysicists to identify key problems and speculate about likely future developments in their areas of expertise. A booklet comprising seventy-three highly varied written responses was then compiled and circulated, and the task of discussing these responses delegated to thirteen invited speakers. Specifically, each speaker was instructed not only to present his own ideas but also to draw attention to or comment upon the views expressed by other respondents, particularly by those who had neither been invited to speak nor were likely to have their ideas discussed by other invited speakers. In my case, the relevant contributions to the booklet would seem to be (in alphabetical order) those by Banks (electrical conductivity), Busse (geodynamo), Green (climate feedback loops), Harrison (core of the Earth), Pochtarev (geomagnetism), Verosub (polarity transitions), Vishik (dynamo problem) and Wilkins (rotation of the Earth). I trust that qualified members of the audience will during the general discussion period bring up all the important questions raised by these respondents. In my allotted twenty minutes, any attempt to do more than list the contributions would be an impertinence to the authors and audience alike.

P.B.Medawar's inspiring collection of articles published under the title "The Art of the Soluble: Creativity and Originality in Science" (Penguin Books, 1967) bears directly on any serious discussion of the strategy of research. Particularly illuminating is Medawar's remark that "no scientist is admired for failing in an attempt to solve problems that lie beyond his competence..... Research is surely the art of the soluble (my italics)...... The spectacle of a scientist locked in combat with the forces of ignorance is not an inspiring one if, in the outcome, the scientist is routed". What I believe are becoming known as "wish lists" have some attractions, but the more difficult "art of the soluble" must be a central consideration in any discussion of key problems for future research, and I take it that the identification of such problems is what this "Quo Vadimus" symposium is supposed to be about.

To conclude this introduction, it is worth recalling that twentieth-century research in geophysics and planetary physics has not only deepened our understanding of the structure, dynamics and evolution of the main constituents of the planetary system, but it has also advanced key areas of classical physics, with insights and discoveries that complement those of modern physics at a high intellectual level. For example, certain basic insights into the behaviour of non-linear systems in many fields of science have their roots in laboratory experiments and theoretical studies in geophysical fluid dynamics. This potential for advancing basic physics should be recognised and exploited in any strategy for planning new work in Earth and planetary sciences. Important work in geophysical fluid dynamics will in the future continue to be motivated by the realization that serious attempts to interpret observations of oceans, atmospheres and magnetic fields, to formulate crucial field experiments, and to provide a satisfactory theoretical underpinning to practical activities such as weather forecasting, will be inseparable from progress with research on dynamic and magnetohydrodynamic processes in rapidly rotating fluids. An important area of fluid dynamics which has direct implications for research on planetary interiors is the study of convection in fluids of high and variable viscosity. The dynamics of tenuous plasmas lies at the heart of magnetospheric studies. "Supercomputers" will facilitate research in all these areas and their availability will be crucial to some of them, such as the magnetohydrodynamics of rotating fluids and the generation of magnetic fields by fluid motions.

Motions in the Interiors of Planets and Variations in their Magnetic Fields

The so-called geomagnetic "geocentric axial dipole" (G.A.D.) hypothesis — that when averaged over a few thousand years the geomagnetic field at the Earth's surface is close in form to that of an imaginary magnetic dipole situated at the geocentre and aligned along the Earth's rotation axis — has been used successfully in geological studies, including the testing in the 1950's of Wegener's proposals concerning continental drift, by palaeomagnetic workers. But more attention must be given to devising palaeomagnetic and other tests of the limitations of the G.A.D. hypothesis. Changes in the geomagnetic field on timescales of core motions, including the tendency for some features of the non-dipole field to migrate westward at a fraction of a degree of longitude per year, undoubtedly smooths non-axisymmetric features of the field to some extent. Complete smoothing, however, is likely to be a much slower process, as became evident in the 1960's when I argued largely on theoretical grounds that departures from axial symmetry in the thermal and mechanical boundary conditions imposed on core motion by various processes, particularly by deep convection in the mantle, should produce distortions in the main geomagnetic field that would average out very slowly indeed, on geological timescales of millions of years. Non-axisymmetric features of the core-mantle interface would also influence the amplitude and other properties of the geomagnetic secular variation and the frequency of polarity reversals, both of which are also amenable to investigation by geomagnetic workers. The IUGG is about to set up a special "Study of the Earth's Deep Interior" (SEDI), recognizing that the time is probably ripe for an interdisciplinary concerted effort to be made to understand in some detail the structure, dynamics and composition of the seventy-five percent of the whole Earth that lies beneath the upper-mantle. The notion of deep intermittent mantle convection interacting strongly with the core could serve as a useful paradigm, to be exploited and if necessary modified later in the light of new knowledge.

The "geocentric axial dipole" hypothesis appeared to gain support when it was found by radioastronomers and space scientists, starting in 1955, that the dipole axes of the planets Jupiter and Saturn nearly coincide with their rotation axes. So it came as a surprise last year when, thanks to the "Voyager" mission of NASA, the present magnetic field of Uranus was found to be highly eccentric, with the dipole axis inclined at 60° to the axis of rotation. The strong challenge presented by this discovery must be taken seriously by theoreticians. Geophysicists accept that strong planetary magnetic fields are due to internal electric currents generated by dynamo action involving the flow of electrically-conducting fluid material. Gyroscopic forces acting on this fluid flow affect the tilt of the magnetic axis, in most cases producing near alignment. A strategy can be formulated for reconciling these general ideas with the Uranus result, without making elaborate ad hoc assumptions about the properties of the planet, but detailed work is needed and the full theoretical implications of the eccentricity of Uranus's magnetism must be assessed.

Fluctuations in the Earth's Rotation

Highly interdisciplinary studies of changes in the Earth's rotation rate and polar motion are now advancing rapidly, thanks to improvements in observational techniques and international cooperation, and the availability of better models of the Earth. Routine determinations of daily values of all three components of the angular momentum of the atmosphere are contributing (a) directly to research on fluctuations in the general circulation of the atmosphere and the extent to which they can account for short-term changes in the Earth's rotation, including the Chandlerian wobble, and (b) indirectly to research on non-meteorological sources of excitation of Earth rotation changes.

Several years ago the theory of angular momentum exchange between the atmosphere and solid Earth was improved and then applied, with the aid of other meteorological and other geophysical data, to the dynamics of the "atmosphere-ocean-solid Earth" system. Short-term changes in the Earth's rotation speed have been shown to be largely of meteorological origin, so that Earth rotation data can be used as a proxy meteorological data set. By exploiting this finding it was established that the global atmospheric circulation fluctuates persistently if irregularly on timescales ranging from about 30 to 80 days. This intraseasonal fluctuation is clearly of potential practical significance in long-range weather forecasting, and it is now being studied intensively by research groups throughout the world seeking to understand underlying dynamical processes. It is not yet clear whether the phenomenon is a manifestation of relaxation oscillations of the Hadley circulation in the Tropics, involving large-scale interactions with the oceans, or alternatively of slow instabilities of mid-latitude flow. Other areas of interest are studies of the direct role of stratosphere in the angular momentum balance and the indirect role of the oceans in modifying the excitation of polar motion.

An important response to these findings was made in 1983 when the IAG of the IUGG organized a special study group (SSG 5-98) charged with promoting further studies of all aspects of atmospheric excitation of changes in the Earth's rotation. The study group has urged meteorologists to forecast atmospheric angular momentum (AAM) changes on timescales of up to about 10 days from the output of reliable global numerical weather prediction (GNWP) models. One practical objective is to provide useful routine forecasts of changes in the length of day (and polar motion) for distribution through the newly-formed International Earth Rotation Service (IERS) of the IUGG and the International Astronomical Union (IAU) to a variety of users, including geodesists concerned with spacecraft navigation and astronomers. AAM forecasts have now been started by three organizations, the U.S. National Meteorological Center, the U.K. Meteorological Office and the European Centre for Medium Range Weather Forecasts. The initial results are encouraging and it would be helpful if other meteorological centres would join in. Further advances in geodesy and geophysics will undoubtedly stem from this new venture, which should also deepen our understanding of the general circulation of the atmosphere and contribute to future improvements in the performance of global numerical weather prediction models.

In the absence of other agencies apparently capable of accounting for the magnitude of the irregular so-called "decade variations" in the length of the day of up to about 5×10^{-3}s, geophysicists accept that they must largely be due to angular momentum exchange between the core and mantle. Concomitant fluctuating torques at the core-mantle interface are produced by time-varying fluid motions in the liquid metallic core. The implied stresses at the core-mantle interface arise as a result of the action of (a) tangential viscous stresses in the Ekman-Hartmann boundary layer, (b) tangential Lorentz forces associated with the interaction of electric currents in the weakly-conducting lower mantle with the magnetic field there, and (c) the action of normal pressure forces on bumps (i.e. departures in shape from axial symmetry) on the core-mantle boundary. The investigation of the relative effectiveness of these agencies is clearly a matter of importance in the study of the structure and dynamics of the Earth's deep interior.

The contribution of viscous stresses is unlikely to be significant except under extreme assumptions about the coefficient of viscosity of the core, and it was proposed (by Bullard) in the 1950's that electromagnetic coupling must be the principal agency. Subsequent refinements in theoretical models of electromagnetic coupling have strengthened the original case for invoking that mechanism, but both qualitative and quantitative difficulties remain, the latter being associated with assumptions concerning the strength of the toroidal part of the geomagnetic field in the outer reaches of the core and the distribution of electrical conductivity in the lower mantle. The idea of topographic coupling was proposed in the 1960's (by the author) on the grounds that the magnitude of the stresses implied by the amplitude and timescale of the decade variations in the length of the day might easily be accounted for if there are bumps on the core mantle boundary of height no greater than about a kilometer and possibly less. Such bumps could easily be maintained by viscous stresses associated with deep convection in the mantle — not a popular idea in the 1960's and 1970's, when mantle convection was generally regarded as being confined to the top 700kms, but now accepted by many geophysicists! How bumps and horizontal temperature variations at the core-mantle interface due to deep mantle convection influence core motions, thereby distorting the Earth's magnetic field, poses important questions in geophysical fluid dynamics, the further investigation of which should be of considerable theoretical and practical significance in the near future.

It is also be of interest to consider whether estimates of topographic coupling can be obtained more or less directly from geophysical data. A method for doing this has recently been proposed by the author and its practical applicability is now being studied by a group consisting of R.Clayton, B.Hager and M.A.Spieth of the California Institute of Technology, C.Voorhies of the NASA Goddard Space Flight Center, and the author. From geomagnetic secular variation data, fields of horizontal motion just below the core-mantle interface are obtained on the basis of a method that exploits Alfvén's frozen magnetic flux theorem plus additional reasonable hypotheses concerning the dynamics of the flow. Horizontal pressure gradients are obtained from these hypothetical velocity fields on the basis of the geostrophic relationship, which should apply in the outer reaches of the core, where the largest ageostrophic term (the Lorentz force) is probably no more than about 10^{-2} times the Coriolis force in magnitude. Gravity and seismic data incorporated in various rheological models of the mantle provide hypothetical topographic maps of the core-mantle interface. The "decade" contribution to changes in the length of day and corresponding changes in the direction of the Earth's rotation axis (polar motion) are obtained from astronomical observations of the Earth's rotation when allowance has been made for tidal effects and short-term contributions due to the atmosphere. First results are encouraging, for they show that for the one epoch studied to date topographic coupling could account for the observed changes in the Earth's rotation both qualitatively and quantitatively, without having to invoke extreme models of core-mantle interface topography and fields of core motions. The reconciliation of these findings with other geophysical work on the structure and dynamics of the core and lower mantle, including certain studies in seismic tomography and geochemistry, will be of importance in the SEDI programme.

Concluding Remarks

As the first speaker in the scientific part of this symposium, I do not have the advantage of hearing other presentations before delivering my own. It will be interesting to see how the discussion progresses and in particular to see what common themes emerge. At a loss for concluding words with useful scientific content, I take refuge in a quotation noticed only the other day whilst walking along a drab corridor during a visit to an underground laboratory of a famous nuclear physics institute. "Earth! My likeness: Though you look impressive and spheric there: I now suspect that is not all". These prophetic words of Whitman are to be found on one of several posters which the U.S. National Academy of Sciences must have distributed widely thirty years ago to draw attention to the International Geophysical Year organized by the IUGG. Let us hope that this "Quo Vadimus" symposium stimulates further great ventures by geophysicists throughout the world, many of whom will be familiar with Piet Hein's exhortation that "problems worthy of attack, prove their worth by hitting back", and also with his warning only by true cooperation can we avoid situations where "our choicest plans have fallen through, our airiest castles tumbled over, because of lines we neatly drew and later stumbled over".

HYDRODYNAMIC COMPLEXITY IN THE EARTH SYSTEM

W. Richard Peltier

Department of Physics, University of Toronto,
Toronto, Ontario, Canada M5S 1A7

Geophysical Fluid Dynamics (GFD), as a recognizably distinct subdiscipline in the geophysically sciences, was probably born in the continuing series of Summer Schools in GFD that began at the Woods Hole Oceanographic Institution over 20 years ago. The goal of these schools was to bring together relatively small groups of gifted graduate students with professional academics working in the areas of astrophysics, atmospheric science, geophysics, oceanography and other areas in which models based upon the concepts of classical hydrodynamics were coming to be seen as central to the understanding of a wide range of dynamical processes. The point of this effort was, and remains, to emphasize the commonality of physical process that underlies the behaviour of such apparently unrelated systems and thereby to stimulate the growth of a new group of theoretical geophysicists whose members are as comfortable in analyzing the behaviour of the infinite Reynold number flows that dominate the general circulation of the planetary atmosphere as they are in developing models of the zero Reynolds number flow associated with convection in the earth's mantle. The goal was, and remains, for example, to educate astrophysicists to understand that the double diffusive processes at work in magnetoconvection in the earth's core (say) have many similarities with the processes that operate in the oceanographically important heat-salt system; and similarly to educate oceanographers to understand that the process of Gulf Stream ring formation has everything to do with the process of occlusion of a frontal baroclinic wave in the atmosphere, etc. The summary of the present state and future promise of this young science, by Raymond Hyde, in the preceding paper of this volume, has provided an interesting view of some of the areas of present research that are liable to be most productive of new insights in the immediate future. By way of this invited response to what Dr. Hyde has written I thought I might amplify somewhat on this same general theme and provide a few further examples of what the future might have in store for those of us who are practitioners of the GFD art. I think I shall begin in the troposphere and work down!

The Planetary Atmosphere

As Dr. Hyde has pointed out, one of the more interesting and novel new lines of recent research in the area of atmospheric dynamics has involved work on the angular momentum budget based upon the interplay between space geodetic measurements of the angular momentum of the solid earth and measurements of the angular momentum of the atmosphere using data from stations of the standard meteorological observing network. Although this work has demonstrated that the exchange of zonal angular momentum between these two components of the earth system is very precisely balanced, and that there is evidence that the Chandler excitation is essentially of meteorological origin, the details of the latter argument do remain somewhat obscure. In this respect I believe it very interesting to note the recent advances that have been achieved in understanding at least one of the mechanisms through which the exchange of angular momentum is mediated. This concerns the influence of what has come to be called "gravity wave drag". Experience with atmospheric general circulation models has shown that a large number of the systematic errors which have plagued such computational models are ameliorated by accounting for the drag exerted on the mean flow by the "breaking" of topographically forced internal waves. This is the same process that Peltier and Clark (1979) have shown to be intimately involved in the genesis of the Chinook, Foehn, and Bora windstorms of North America and Europe. No detailed attempt has yet been made to understand the impact of this "unresolved" dynamical processes in the general circulation upon the excitation of the Chandler wobble but such work is needed and will likely provide new insights into the solid earth-atmosphere interaction. This process should be seen as a "topside" analogue of the same topographic core-mantle coupling mechanism discussed by Hyde in the preceding paper.

Also in the area of atmospheric dynamics, it is quite clear that the next decade will see very significant increases, because of the development of ever more powerful computer systems, in our ability to explicitly resolve important dynamical processes which are unrepresented in the current generation of dynamical models. Those which will first begin to come into focus as resolving power increases will be the frontal structures that are characteristic of synoptic scale mid-latitude cyclones. Perhaps we will also begin to see in our models the genesis of the frontal cyclones that formed the heart of the dynamical meteorology invented by Bjerknes and other members of the Norwegian school whose views were subsequently eclipsed by the discovery by Charney and Eady of the process of baroclinic instability. We may well thereby discover that fronts are not only caused by the nonlinear evolution of long-wavelength Charney-Eady waves but that they are also themselves responsible for the origin of shorter horizontal wavelength, shallow, frontal cyclonic disturbances - the frontal waves of Bjerknes (Moore and Peltier 1987).

The Mantle of the Earth

One of the most interesting puzzles in the "solid earth" geophysical sciences remains an array of issues concerning the way in which the thermal convection process that drives continental drift is energized and the way in which the associated surface velocity field is caused to assume the "plate-tectonic form" that was first deduced almost twenty years ago. Although the results from seismic tomographic imaging of the depth dependent lateral heterogeneity of body wave velocities, and thus temperature, have enabled rather satisfying explanations to be provided of the large scale a-spherical geoid (Hager 1984) and the poloidal part of the surface velocity field itself (Forte and Peltier 1987), we still have no satisfactory explanation of why the surface kinetic energy is almost precisely equipartitioned between the poloidal (ridges and trenches) and the toroidal (transform faults) components of the flow (e.g. Peltier 1985). Until we have designed a convection model that is capable of explaining this equipartition we will be in no position to employ such models to infer the deep mantle viscosity profile and therefore in no position to understand whatever differences may exist between the viscosity preferred by the long-timescale process of mantle convection and that inferred on the basis of the short timescale process of postglacial rebound. This issue is clearly rather crucial in our attempts to better understand mantle rheology.

New seismic data that constrain the topographic relief on the core mantle boundary are expected to provide an extremely useful further constraint on the nature of the convection process and the future should see considerable improvement in the presently low accuracy of this inference. Other issues of interest regarding the nature of the mantle convection process concern the impact on the radial structure of the circulation of the solid-solid phase transformations that bracket the transition zone at 420 km and 670 km depth, and the extent to which temperature and pressure dependence of the viscosity may exert a controlling influence upon the heat transfer and therefore on the planetary thermal history. Advances in supercomputer technology will in future enable us to begin to construct high resolution three dimensional spherical models that incorporate all of these influences and with them we will be able to begin to explore the detailed properties of the thermally chaotic motions that control the thermal evolution of the planet.

The Earth's Core

Although it is the mantle convective circulation that governs the rate of planetary cooling, it is the response of the electrically conducting core to this cooling that is responsible for the generation of the planet's magnetic field. As discussed above, research on the nature of the mantle convection process has entered a phase in which at least the concept is no longer at issue and work has come to be directed towards understanding the details of the connections between energetics and kinematics. Insofar as the theory of mhd processes in the core is concerned, however, I think it fair to say that the situation remains one in which a number of rather fundamental issues are still "hotly" disputed. One important ongoing debate concerns the basic nature of the force balance that governs the operation of the geodynamo. Here an important dispute concerns the relative viability of models of so-called Taylor type for which core-mantle coupling is irrelevant and those of so-called z-type in which it is crucial. No detailed mhd models currently exist that can be employed to directly investigate the influence of topographic effects on the dynamo process itself. In fact we are still unable to explain why the planetary magnetic field has the dipole moment it has, and we have only the crudest possible dynamical systems models, derivative, for example, of the circuit equations for the shunted disc dynamo, to guide us in understanding the magnetic reversal phenomenon. In many ways these issues still constitute a most formidable array of theoretical problems - and certainly the most fundamental - in all of the geophysical sciences.

Summary

Over the course of the next decade it seems clear that our science will be evolving in a way that stresses integration of the subdisciplines. This evolution will clearly be required if we are to seriously address the pressing issues of Global Change which are not only on the future horizon but actually at our door-step now. We need to begin to develop an integrated view of the coupled evolution of the whole earth system, including its atmospheric, oceanographic, cryospheric, biospheric, and solid earth components. It is clear that the multidiscipline of geophysical fluid dynamics will play an important role in guiding this development.

References

Forte, A.M. and W.R. Peltier, Plate tectonics and a-spherical earth structure: the importance of poloidal-toroidal coupling, J. Geophys. Res., 92, 3645-3679, 1987.

Hager, B.H., Subducted slabs and the geoid: Constraints on mantle rheology and flow, J.Geophys. Res., 89, 6003-6015, 1984.

Moore, G.W.K. and W.R. Peltier, Cyclogenesis in Frontal Zones, J. Atmos. Sci., 44, 384-409, 1987.

Peltier, W.R. and T.L. Clark, The evolution and stability of finite amplitude mountain waves. Part II: Surface wave drag and severe downslope windstorms. J. Atmos. Sci., 36, 1498-1529, 1979.

Peltier, W.R., Mantle convection and viscoelasticity, Ann. Rev. Fluid Mech., 17, 461-608, 1985.

CLIMATE

George C. Reid

Aeronomy Laboratory
National Oceanic and Atmospheric Administration
Boulder, Colorado, U.S.A.

Introduction

There is fairly general agreement that man's activities will cause significant changes in the earth's climate on both regional and global scales during the next century. Clearly we need to be able to predict with some degree of precision the nature and magnitude of these changes, particularly at the regional and local level, where their impact is most clearly felt. In order to achieve this predictive capability, we must first understand the natural mechanisms that drive climate change, and the likely range of variability that these will produce over the next few decades.

The earth is a rapidly rotating planet largely covered with liquid water, and the general circulation of the atmosphere as we know it is the response to the latitudinal gradient in surface heating of such a planet. The earth's rotation rate and the distribution of surface water are unlikely to change appreciably within the time scales of interest, but variations in the distribution and intensity of surface heating have probably been a major cause of climate change in the past, and are likely to remain so in the near future. Such changes can arise from a wide variety of causes, both internal and external, and understanding the detailed response of climate to all conceivable kinds of forcing is a formidable challenge that will perhaps never be met. If we understood the response to a relatively simple change in external forcing, such as a change in the luminosity of the sun, we would have gone a long way towards understanding the geographical linkages and feedbacks that exist in the system. Since evidence is now accumulating that the sun's luminosity is in fact variable at levels of the order of 0.1 percent on the decadal time scale, and possibly of several tenths of a percent on the interdecadal time scale, this question is a timely one of great intrinsic scientific and practical interest.

General circulation models of the atmosphere have predicted a high degree of sensitivity to changes in the solar "constant", yet the paleoclimatic evidence clearly shows that the surface temperature has not gone through extremes, even when the sun was significantly fainter than today, in the early days of biological evolution. Paradoxically, though, the advances and retreats of the Pleistocene ice sheets have apparently been driven by quite small changes in the distribution of solar radiation over the earth's surface caused by cyclical changes in the earth's orbit. These Milankovitch variations are too slow to be of concern in the near future, but variations in the sun's intrinsic luminosity may occur on much faster time scales.

Key Questions

How Does the Earth's Radiation Budget Respond to Changes in Solar Radiation?

The hydrological cycle is of vital importance here. Changes in evaporation from the oceans lead to changes in tropospheric water vapor, which largely controls the "greenhouse" properties of the atmosphere, leading to a positive feedback situation. Cloud cover is also likely to be increased in certain regions, giving rise to major perturbations of the radiation budget through an increase in the planetary albedo and through a reduction in the longwave radiation to space by high clouds. The increased albedo tends to cool the earth, while the reduction in escaping longwave radiation tends to warm the earth. The extent to which these competing effects cancel each other is a topic of much current research, but is not yet resolved. Regionally, the largest effects would be expected in the tropical regions of intense convection: the "maritime continent" of southeast Asia, the Amazon basin of South America, and the equatorial regions of Africa.

How Does the General Circulation of the Atmosphere Respond to Changes in Solar Radiation?

This question is intimately linked to the previous one. It is also bound up with the response of the oceans, since variations in sea-surface temperatures, particularly in the tropics, are known to have a profound effect on the general circulation. Since the winds themselves partially determine sea-surface temperatures through evaporation and upwelling, the feedback aspect is strong.

On the regional scale, changes in solar radiation would be expected to have a significant impact on the intensity of the Hadley and Walker circulations of the tropics and on the strength of all monsoon circulations, including the vitally important summer and winter monsoons of southern Asia. The drought-prone regions of subtropical Africa are also strongly influenced by these monsoon circulations, as well as by sea-surface temperature variations in the tropical Atlantic and by the intensity of the local Hadley circulations. The strengths of the subtropical high-pressure belts are likely to change, and with them the strength of the trade winds. In the Pacific there are likely to be corresponding changes in the strength of the zonal Walker circulation, and in the intensity of the periodic El Niño events that have devastated the South American coast at times. In both the Atlantic and the Pacific, the frequency and intensity of tropical storms and hurricanes are likely to change in ways that are currently unpredictable.

These changes in the tropics and subtropics will be accompanied by changes at mid-latitudes and in the polar regions whose nature is not clear, and whose magnitude is unpredictable. Some of these changes will be driven by the change in local heating, and some by the change in the export of heat from lower latitudes brought about by the changes in the tropical circulation.

Understanding the nature and magnitude of these regional climate changes is a major challenge. Much more data are needed from the vast uninhabited ocean areas of the earth, and satellite remote sensing has a major role to play. The record of past climate variability also has much to tell us.

To What Extent Could Solar Variability Induce Changes in Atmospheric Composition That Might Lead to Climate Change?

Examples might be changes in the balance of atmospheric versus oceanic carbon dioxide or changes in the biomass. That such changes have occurred in the past is clear from the paleoclimatic record: the global warmth of the Cretaceous was probably accompanied by an increase of global biomass over present-day levels, while chemical analysis of ice cores from Greenland and Antarctica has shown that atmospheric CO_2 concentrations were significantly reduced during the Pleistocene glacial periods. In the case of the Pleistocene, current thinking suggests that the effect was a consequence of the redistribution of solar heating through the Milankovitch orbital variations rather than a result of intrinsic solar variability, but the Cretaceous warmth is more enigmatic. Clearly a more detailed study of the existing paleoclimatic evidence is crucial to an understanding of this problem, in which the potential for strong feedback is also evident.

How Important are the Present-Day Polar Ice Sheets (Antarctica and Greenland) in Determining our Present Climate?

The permanent polar ice sheets affect global climate in at least two distinct ways. They increase the planet's albedo, thereby lowering the mean temperature and tending to stabilize themselves, while their existence tends to strengthen the temperature gradient between equator and poles that largely drives the general circulation of the atmosphere. These facts are widely recognized, but their consequences for regional climate are not well understood. Changes in solar radiation must have some impact on the properties and geographical extent of the ice sheets, and through them on global and regional climate. For example, an increase in solar radiation might lead to destabilization of the oceanic edges of the ice, an increase in the rate of calving of icebergs, and a cooling of the surface waters. In the case of the Greenland ice sheet, this could have important consequences for the climate of western Europe. The overall problem of the interaction of the cryosphere, the oceans, and the atmosphere is an extremely complex one that poses a formidable challenge to current modeling capability.

COMMENTS ON GEORGE REID'S "QUO VADIMUS" CONTRIBUTION "CLIMATE"

Paul J. Crutzen
Max-Planck-Institute for Chemistry,
Airchemistry Dept.
P.O. Box 3060, D-6500 Mainz, F.G.R.

It is fascinating to observe George Reid gaze into his crystal ball. He has identified some of the major problems in climate research. Especially the existence of some strong positive feedbacks in the global climate system and the expected impact of human activities on climate through the emissions of greenhouse gases, such as CO_2, CH_4, N_2O, $CFCl_3$ and CF_2Cl_2, may produce some major climatic perturbations with the next decades to centuries.

If major changes were indeed to occur, then one can be pessimistic about Mankind's ability to adjust in a controllable way, the more since climate changes could be very rapid and even of an oscillatory nature, as was observed during the exit from the last glacial period. Hopefully, our scientific fears will prove to be unfounded.

George Reid does not mention the feedbacks which exist between climate, ocean dynamics and biological responses. These issues will be at the center of the World Climate Research Program and the International Geosphere Biosphere Program on Global Change, the most ambitious global environmental research efforts that have ever been planned and that will be carried out in the next decades. I expect some considerable progress in our understanding of the working of the global environment from these programs. It is precisely, as George Reid recommends, aimed at developing our ability to predict climate over the next century, including biospheric interactions. For instance, if a substantial fraction of the large amount of organic carbon that is locked up in subarctic ecosystems were released as CH_4, positive feedback effects on climate and global atmospheric chemistry could be very large indeed.

George Reid says little about the upper atmosphere, a field of research to which he has contributed in such admirable and diverse ways. Recently, we have again been surprised by Nature's response to the activities of only a small fraction of the human population. The sudden appearance of the Antarctic "ozone hole" during each springtime with a total ozone loss by a factor of two in less than a decade, has been really dramatic and at the same time depressing for us scientists, because it shows us how little we know and how far we may still be away from real predictive capabilities. The stratosphere is most likely of enormous importance for climate (after all, most convection stops at the tropopause). Furthermore, the strong increase of ozone concentrations with height and ozone heating by the absorption of solar ultraviolet radiation leads to the stratospheric temperature inversion that, in turn, allows for high ozone concentrations in the lower stratosphere, hence providing for a positive coupling. The response of tropopause height and temperatures to anthropogenic disturbances is, in my opinion, a very important scientific problem with potentially major implications for stratospheric water vapor (and, therefore, ozone chemistry) and maybe for climate in general.

RESEARCH IN CLIMATE SCIENCE
Response to paper by George C. Reid

Stephen H. Schneider*

National Center for Atmospheric Research**

George Reid has provided a valuable discussion of a number of aspects of climate science research to be emphasized in the next few decades. I have no quarrel with his views but feel some additions and an overall perspective may be helpful, and I thank the editors of the Quo Vadimus volume for soliciting my opinions.

I shall briefly present my views of the needs for climate research in the next several decades in a brief outline of the search for a validated theory of climate in nine specific climate problem areas that span the space and time scales of important climatic variations.

The "Goldilocks" Problem

It has been quipped that the climate of Venus is too hot, Mars too cold, and the earth just right for life [for example, Margulis, 1986, or Kasting et al., 1988]. Although a number of plausible theories and calculations have been advanced to suggest the very different climatic fates of the terrestrial planets, interesting puzzles still remain in comparative planetary climatology. For example, the spacecraft evidence for former large amounts of water vapor in the atmosphere on Mars seems clear, but the explanation of its disappearance remains an unsolved issue.

---

*Any opinions, findings, conclusions or recommendations expressed in this editorial are those of the author and do not necessarily reflect the views of the National Science Foundation.

**The National Center for Atmospheric Research is sponsored by the National Science Foundation.

Geophysical Monograph 60
IUGG Volume 10
©American Geophysical Union

The Faint Early Sun "Paradox"

Most conventional astrophysical theory suggests that the luminosity of the sun was some 30% less 4 billion years ago than at present. This has led to what some have called a "paradox," whereby the earliest rock evidence suggests that some 3-1/2 to 4 billion years ago liquid water was present on earth as well as fossilized bacteria in rocks. The presence of sedimentary rocks and life itself suggests temperatures probably in the $0°$ to $40°C$ range. A paradox arises if certain climatic models are subjected to 30% decrease in solar luminosity, since they imply a frozen earth. Early attempts to reconcile this paradox were offered by Sagan and Mullen [1972], postulating a super greenhouse effect of methane and ammonia. Later it was suggested that primordial CO_2 associated with enhanced volcanism could solve the faint early sun paradox [Owen et al., 1979]. The largest uncertainty, however, may well be determining from the very fragmentary rock record precisely what the climate was. Were the patches of fossilized life and sedimentary rock representative of a wide distribution of favorable conditions around earth or, rather, localized exceptions? Was the planet very cold on average and life restricted to geothermal hot spots or equatorial areas or, conversely, the planet very hot on average with life holding on in the polar regions? Finally, recent astrophysical theories suggest a solar evolution model involving substantial mass loss would run the faint early sun paradox in reverse, since such mass loss models predict a solar constant perhaps 30% higher in the Archean age than present! Much remains to be learned about this interesting but dimly perceived period of earth history.

The Coevolution of Climate and Life

Perhaps more interesting than the faint early sun paradox is the consequence that necessarily follows: if the faint early sun paradox is to be "solved" by a super greenhouse effect, then the obvious problem switches from maintaining warmth at the faint young sun to preventing overheating as the solar luminosity increased. Most authors suggest that the solution to this problem involves the negative feedback process whereby carbon dioxide is removed from the earth's atmosphere by the weathering of silicates and other minerals and their deposition as carbonate rocks. Debate has been intense over the relative importance of biological processes [Lovelock, 1988] versus inorganic weathering [for example, Walker et al., 1981; Kasting et al., 1988] in the CO_2 removal [for example, see review by Schneider, 1990]. As continents rose and moved, atmospheric composition radically changed (for example, oxygen increased and CO_2 decreased, presumably) and the sun heated up, life evolved and diversified, modifying the physical and chemical environment in ways that could feed back on the very climate that determined the ecological niches of individual species. Therefore, a study of the coevolution of climate and life between the end of the Archean and the Cambrian covers a time when considerably more geologic record can be brought to bear to constrain our unbridled scientific imaginations.

Mid-Cretaceous Warmth "Paradox"

A multiplicity of proxy indicators of climate suggest that the mid-Cretaceous, some 100 million years ago, was a time of extraordinary warmth -- mean global temperatures at least $5°$ and up to $15°C$ warmer than today. Paleobotanists argue that fossilized specimens found in mid-continental areas in the northern hemisphere would indicate the absence of severe frost, even in mid-winter in mid-continent. Climatic model simulations using mid-Cretaceous paleogeography and assuming that the seasonal cycle of solar radiation is the same as today are unable to produce mid-continent, mid-winter temperatures in the northern latitudes without some substantial occurrence of frost [Schneider et al., 1985]. Even when we allow Arctic Ocean temperatures in the model to be as warm as $20°C$ -- permitting a year-round above-freezing Arctic seaboard which presumably could have sustained duckbill dinosaurs -- it still was not possible to prevent nighttime infrared radiative cooling from dropping mid-continental surface temperatures below freezing on a number of different days in the winter. Perhaps one solution to the mid-Cretaceous, mid-winter warmth paradox would be a substantially enhanced greenhouse effect, whereby perhaps a factor of 10 more carbon dioxide may have been present [e.g., Budyko and Ronov, 1980 or Berner et al., 1983]. Resolution of these difficulties should be part of the scientific agenda for the next several decades.

Cenozoic Transition

It is widely believed (based on oxygen isotope analyses of both surface and benthic foraminifera) that the Cenozoic (65 million years ago - present) era saw a gradual deterioration of the climate, whereby permanent ice took hold at the poles and a very large regional differentiation in biotic zones developed in response to the $5°$ to $15°C$ cooling in global average planetary temperature. A number of plausible theories have been advanced, including reductions in sea floor spreading rates leading to lowered CO_2 amounts in the atmosphere, changes in paleogeography and so forth. Nevertheless, a validated quantitative theory of Cenozoic climate transition remains to be explored in considerably more depth over the next several decades.

The 100,000-Year Cycle (The Ice Age/Interglacial Cycles)

By 2-1/2 million years ago, the beginning of the Pleistocene, the geologic record suggests that rapidly fluctuating ice age/interglacial stages replaced the slow but seemingly smoother Cenozoic transition toward permanent ice-covered poles. These ice ages grew in magnitude and appeared to have substantial variability at frequencies of around 20,000 years, 40,000 years, and 400,000 years, at least until about 1,000,000 years ago. The past million years of the late Pleistocene have seen the largest amplitude fluctuations in ice age/interglacial cycles, but for this period the dominant cycle is the 100,000-year cycle. Some people have suggested that the 19/23,000-year peak is a response to the precession of the equinoxes and the 41,000-year peak in variance is a response to the obliquity cycle of the earth. These orbital variations are caused by the tug of the gravities of other planets. There is some small solar radiative forcing from orbital variations at 100,000 years but not nearly enough to explain the dominance of the 100,000-year power in the past million year records -- unless one invokes some major positive climatic feedback that seemingly would imply climate sensitivity much greater than that exhibited by the climate system in response to the seasonal cycle or other known forcings. Some have suggested that the 100,000-year power in the recent record is an internal, non-linear response of the climatic system, including atmosphere, oceans, ice and bedrock to the solar insolation forcing at precession and obliquity frequencies [for example, Le Treut and Ghil, 1983]. Others think the 100,000-year cycle might be an amplified direct response to the so-called Milankovich forcing [Imbrie and Imbrie, 1979]. Others invoke biogeochemical cycle feedbacks among climate changes and trace greenhouse gases [Lorius et al., 1990]. In any case, it is a major unsolved scientific question as to why the 100,000 year power that developed so strongly only in the past million years is a "forced" or "free" oscillation of the climatic system or some combination of both.

Abrupt Climatic Surprises

A number of abrupt events in climate have occurred and have attracted scientific interest. By abrupt, we mean the occurrence of the change is as fast or faster than that which can be resolved by the proxy records marking its occurrence. For example, the Cretaceous-Tertiary extinctions of many species of phytoplankton and all large animals has been attributed to extraterrestrial asteroid impacts [e.g., Alvarez et al., 1980], terrestrial events like volcanic eruptions [Officer et al., 1987; Stothers et al., 1986], or other factors. Other abrupt climatic surprises include the rapid reglaciation of at least the North Atlantic sector of the northern hemisphere about 10,800 years ago after it appeared the ice age had ended [for example, see Berger and Labeyrie, 1987]. This Younger Dryas epoch lasted several hundred years and may very well have involved a very rapid dramatic reversal in the deep circulation of the oceans with worldwide implications. Investigation of the nature and causes of such abrupt events will certainly be a major part of climate system research in the next several decades.

Greenhouse Effect: Transient Response

Much attention has been given to the equilibrium response of climatic models driven by a forced doubling of carbon dioxide. Changes in temperature, rainfall, soil moisture and so forth are typical outputs of such models and have been used to estimate potential environmental and societal impacts of human-induced climate change on agriculture, water supplies, forests, sea level and so forth. However, it has been recognized over the past decade [for example, Schneider and Thompson, 1981; Bryan et al., 1982] that the evolution of trace gas forcing is not a simple step function increase in CO_2 from which an equilibrium climate is compared to a control climate. Rather, time evolving increase in trace gas greenhouse forcing suggests that different parts of the climatic system, particularly different parts of the oceans which mix vertically at different rates, would warm differentially over time. (Indeed, that is what preliminary, coupled, three-dimensional atmosphere/ocean general circulation model simulations suggest [for example, see Washington and Meehl, 1989 or Stouffer et al., 1989].) This implies that the horizontal temperature gradient could evolve in time differently from that in equilibrium, thereby suggesting that regional climatic anomalies during the transient phase of climate warming over the next several decades could be of different character than that inferred from equilibrium calculations. It is essential for both the early detection of greenhouse effect signals and the social response to the prospect of increasing trace gases that the transient response of the coupled atmosphere, ocean and sea ice systems and their interaction with soil moisture and other land surface processes be rapidly and carefully studied over the next few decades. Moreover, the evolution of transient temperature change in the ocean will help to guide estimates of the rates of thermal expansion of the oceans which is critical to evaluating the rate of sea level rise on a global basis and its potential implications for coastal flooding during storm surge events or at other times [e.g., Wigley and Raper, 1987].

El Niño and Blocking

At the shortest time scales of climate change, a number of phenomena have been observed to be of major importance to biospheric activity. The southern oscillation and its related "El Niño" effect [Rasmusson and Hall, 1983] are well known to cause radical alterations in the rainfall and temperature regimes across the equatorial Pacific. They are also believed to be involved in the propagation of climatic anomalies to other parts of the world ranging from Africa and Asia to North America. Blocking, which can be thought of as the persistent anomaly of a highly meridional circumpolar vortex, can lead to unusually long periods of anomalous climatic patterns locked into regional positions. These, too, can have substantial impacts on water supplies, agriculture and human and animal health. The causes of these phenomena and our capacity to model their frequency or timing is of major intellectual and social significance because of the substantial impact of regional climatic anomalies on society. Therefore, it is likely that such events will continue to receive research efforts over the next several decades.

So far, I have tried to list, in order of decreasing time scale, a number of "climate problems" that should be a focus of research in the next several decades. Let me conclude with some connected, cross-cutting theoretical issues.

Climatic Determinism

One of the most interesting theoretical debates in climatology is whether the rich spectrum of observed climatic variations are externally forced (for example, by the seasonal cycle of solar radiation) or internally generated by exchanges of mass, momentum and energy among the climatic subsystems (for example, the southern oscillation). In particular, climatic modelers have argued whether the climatic system exhibits deterministic, stochastic or chaotic behavior (or some combination) as it evolves. For example, is the 100,000-year cycle mentioned earlier a forced response in which internal climatic feedback processes amplify a weak orbital signal; is it one possible signature of a complex dynamical system or a principal free oscillation of a system exhibiting stochastic behavior? Another example is the few tenths of a degree C drifts in the decadal trend of temperatures of the planet. Are they externally forced by solar variations or volcanic dust, or SO_2 [Charlson et al., 1987] stochastically driven by the random nature of many unpredictable storm systems being integrated by the slowly responding oceans [Hasselmann, 1976], or the manifestations of a chaotic attractor resulting from the complex internal dynamical nature of the coupled atmosphere, ocean, cryosphere, land, biosphere system [for example, Lorenz, 1968]? Solving this interesting theoretical issue will involve a hierarchy of coupled climate subsystem models, observations to suggest new formulations and to validate existing models, and further interactions among physically and mathematically oriented researchers.

Sensitivity vs. Stability

Two of the important methodological components of climate modeling research deal with the evaluation of the sensitivity of the climate to a given external forcing, most commonly, increase in trace greenhouse gases or changes in solar radiation associated with orbital element variations. For example, typical state-of-the-art, three-dimensional general circulation models suggest that in equilibrium a doubling of atmospheric carbon dioxide would increase global average surface temperature something between $2°$ and $5°C$ [IPCC, 1990]. Stability is another property of the climatic system that is explored through climatic models. Since the sensitivity of the climate to external forcing can change as a function of the mean state of the climate, some modelers have looked to see if multiple equilibria could be found in their climatic models. For example, Budyko, 1969, and Sellers, 1969, suggested that the feedback of increased snow and ice with planetary cooling would increase the albedo, thereby setting off an instability that would drive the climate to an ice-covered earth from which the earth would not recover, even if the solar constant were restored to its present value. This measure of climate stability was extended to variations in initial values by Schneider and Gal-Chen [1973], who found that initial temperature changes from anywhere above present equilibrium and down to a threshold value below present would always reproduce today's equilibrium climate. However, any initial temperature distribution below this threshold, even with present day external forcing, produced the ice-covered earth. These very simple models merely are mentioned to provide examples of methods for determining stability in climate models. Sensitivity and stability characteristics of more complex models will differ, and the increasing trend toward coupling of climatic subsystems (including atmosphere, ocean, ice, land and biospheric subcomponents) will undoubtedly produce models whose sensitivity and stability properties will change.

Observational Systems

Finally, it is critical to the development of improved formulation of climatic models, the validation of models, and the basic understanding of climate system behavior to have increasingly sophisticated empirical data on both modern and paleo climates. Modern climate data sets should include not only traditional aerologic variables but also must be extended to hydrologic processes that include soil moisture and runoff, biospheric phenomena such as net primary productivity of forests or phytoplankton and sources and sinks of radiatively important constituents, and should include extent and variability of sea ice and glaciers. Satellites are an essential component of the new data needs, but air, sea and ground truth are all required to calibrate remote sensing instruments and to help squeeze the most scientific content from the space data. Comprehensive and readily available data sets in common formats should be archived and will help with the increasingly interdisciplinary nature of climate change research as it relates to the development of the International Geosphere Biosphere (IGBP) or Global Change Program [e.g., NAS, 1988]. In addition to modern data sets, proxies of climatic and biological change over time are critical since large changes, particularly over the past 20,000 years since the last glacial maximum, provide the backdrop against which to calibrate our understanding and predictive capacity for the future. Of particular importance are ice cores, which have now been extended backward to one glacial cycle [e.g., Barnola et al., 1987] and should be extended over two cycles where possible. These would provide exciting data on climatic and chemical variability. Ocean and lake cores will continue to provide additional proxies [e.g., Martinson et al., 1987; COHMAP, 1988], and it is likely that these will undergo increased cross-calibration with radiometrically dated land and oceanic records over the next few decades.

In summary, the climate sciences have evolved from the primarily multidisciplinary mode in which various disciplines produced their respective subcomponents of the problem in relative isolation to a new phase of interdisciplinary research, in which workers from many fields are beginning to combine their knowledge, skills, submodels and observations in order to address the broader-scale problems of global change that will be critical for scientific study into the next few decades.

References

Alvarez, L. W., W. Alvarez, F. Asaro, and H. V. Michel, 1980: Extraterrestrial cause for the Cretaceous-Tertiary extinction, Science, 208, 1095-1108.

Barnola, J. M., D. Raynaud, Y. S. Korotkevich and C. Lorius, 1987: Vostok ice core provides 160,000-year record of atmospheric CO_2, Nature, 329, 408-414.

Berger, W. H. and L. Labeyrie (eds.), 1987: Abrupt Climatic Change, NATO/NSF Workshop, Biviers (Grenoble), Reidel, Dordrecht, 399-417.

Berner, R. A., A. C. Lasaga, and R. M. Garrels, 1983: The carbonate-silicate geochemical cycle and its effect on atmospheric carbon dioxide over the past 100 million years, Amer. J. Sci., 283, 641-683.

Bryan, K., F. G. Komro, S. Manabe, and M. J. Spelman, 1982: Transient climate response to increasing atmospheric carbon dioxide, Science, 215, 56-58.

Budyko, M. I., 1969: The effect of solar radiation variations on the climate of the earth, Tellus, 21, 611-619.

Budyko, M. I. and A. B. Ronov, 1980: Chemical evolution of the atmosphere in the Phanerozoic, Geochemistry International 1979, 16, 1-9.

Charlson, R. J., J. Lovelock, M. O. Andreae, and S. G. Warren, 1987: Oceanic phytoplankton, atmospheric sulfur, cloud albedo and climate, Nature, 326, 655-661.

COHMAP Members (P. M. Anderson et al.), 1988: Climatic changes of the last 18,000 years: Observations and model simulations, Science, 241, 1043-1052.

Hasselmann, K., 1976: Stochastic climate models. Part I. Theory, Tellus XXVIII, 6, 473-485.

Imbrie, J. and K. P. Imbrie, 1979: Ice Ages: Solving the Mystery, Short Hills, N.J., Enslow.

Intergovernmental Panel on Climate Change (IPCC), June 1990, Scientific Assessment of Climate Change, Geneva, World Meteorological Organization.

Kasting, J. F., O. B. Toon, and J. B. Pollack, 1988: How climate evolved on the terrestrial planets, Sci. Amer., February, 90-97.

Le Treut, H. and M. Ghil, 1983: Orbital forcing, climate interactions, and glaciation cycles, J. Geophys. Res., 88, 5267-5190.

Lorenz, E. N., 1968: Climate determinism. Meteorol. Monogr., 8, No. 20, 1-3.

Lorius, C., J. Jouzel, D. Raynaud, J. Hansen, and H. LeTreut, 1990: The ice-core record: Climate sensitivity and future greenhouse warming, Nature, 347, 139-145.

Lovelock, J., 1988: The Ages of Gaia, W. W. Oxford University Press, Oxford, 252 pp.

Margulis, L., 1986: NOVA program, "Goddess of the Earth," WGBH, Boston.

Martinson, D. G., N. G. Pisias, J. D. Hays, J. Imbrie, T. C. Moore, Jr., and N. J. Shackleton, 1987: Age dating and the orbital theory of the Ice Ages: Development of a high-resolution 0-300,000-year chronostratigraphy, Quat. Res., 27, 1-29.

National Academy of Sciences, 1988: Toward an understanding of global change: Initial priorities for U.S. contributions to the International Geosphere-Biosphere Program, Washington, D.C.

Officer, C. B., A. Hallam, C. L. Drake, and J. D. Devine, 1987: Late Cretaceous and paroxysmal Cretaceous/Tertiary extinctions, Nature, 326, 143-149.

Owen, T., R. D. Cess, and V. Ramanathan, 1979: Enhanced CO_2 greenhouse to compensate for reduced solar luminosity on early earth, Nature, 277, 640-641.

Rasmusson, E. M. and J. M. Hall, 1983: El Nino, Pacific Ocean event of 1982-1983, Weatherwise, 36, 167-175.

Sagan, C. and G. Mullen, 1972: Earth and Mars: Evolution of atmospheres and temperatures, Science, 177, 52-56.

Schneider, S. H., 1990: Debating Gaia, Environment, 32, No. 4, 5-9, 29-32.

Schneider, S. H. and T. Gal-Chen, 1973: Numerical experiments in climate stability, J. Geophys. Res., 78, 6182-6194.

Schneider, S. H. and S. L. Thompson, 1981: Atmospheric CO_2 and climate: Importance of the transient response, J. Geophys. Res., 86, 3135-3147.

Schneider, S. H., S. L. Thompson, and E. Barron, 1985: Mid-Cretaceous continental surface temperatures: Are high CO_2 concentrations needed to simulate above-freezing winter conditions?. In The Carbon Cycle and Atmospheric CO_2: Natural Variations Archean to Present (E. T. Sundquist and W. S. Broecker, eds.), American Geophysical Union, Washington, D.C., 554-559.

Sellers, W. D., 1969: A global climatic model based on the energy balance of the earth-atmosphere system, J. Appl. Meterology, 8, 392-400.

Stothers, R. B., J. A. Wolff, S. Self, and M. R. Rampino, 1986: Basaltic fissure eruptions, plume heights, and atmospheric aerosols, Geophys. Res. Lett., 13, No. 8, 725-758.

Stouffer, R. J., S. Manabe, and K. Bryan, 1989: Interhemispheric asymmetry in climate response to a gradual increase of atmospheric CO_2, Nature, 342, 660-662.

Walker, J. C. G., P. B. Hays, and J. F. Kasting, 1981: A negative feedback mechanism for the long-term stabilization of earth's surface temperature, J. Geophys. Res., 86, 9776-9782.

Washington, W. M. and G. A. Meehl, 1989: Climate sensitivity due to increased CO_2: Experiments with a coupled atmosphere and ocean general circulation model, Climate Dynamics, 4, 1-38.

Wigley, T. M. L. and S. C. B. Raper, 1987: Thermal expansion of sea water associated with global warming, Nature, 330, 127-131.

SCALE INTERACTIONS

John T. Snow
Department of Earth and Atmospheric Sciences
Purdue University
West Lafayette, Indiana

1. Background

Since the time of the first world war, investigation of synoptic processes has been a major focus of atmospheric research. These are the physical processes that drive the continuously evolving pattern of high and low pressure centers and attendant frontal boundaries that are to be seen on continental-scale weather maps. This effort has been motivated both by a spirit of scientific inquiry and by a desire to improve operational weather forecasting by national meteorological services. These national services in turn have supported the development of a global observational network that provides the data required for both operational and research purposes. As a consequence of this research, there now exists a reasonable physical understanding of many of the phenomena found at this synoptic scale. This understanding is reflected in the numerical weather forecast models used by the national services. These have shown significant skill in predicting the evolution of synoptic-scale features for periods extending out to five days.

It was recognized very early that there are strong interactions between synoptic features and local weather events (now commonly referred to as mesoscale phenomena) that lead to the transfer of heat, moisture, and momentum. The global observing network provides data that is sufficient to monitor the evolution of synoptic features. However, with conventional techniques it has not been practical to routinely observe the embedded mesoscale phenomena in sufficient temporal or spatial detail to produce the data base needed for detailed analysis. Without such analyses, it has proved difficult to quantify the interactions, so progress in puzzling out the physical mechanisms coupling the two scales has been very slow. This has been especially true of those processes wherein mesoscale phenomena modify the larger scale environment.

A practical consequence is that in the development of the large numerical forecast models, interactions between scales have had to be handled through semi-empirical parameterization. This appears to be one of the major constraints limiting many of the current numerical models to forecast periods of about 5 days. Also, the actual prediction of local weather retains a high level of subjectivity. Future local conditions are usually inferred from the forecast synoptic pattern by a human weather forecaster, often on the basis of experience rather than physical insight. Improved physical understanding into the nature of the whole spectrum of scale interactions is essential to improving the large numerical models and ultimately the forecasting of local weather events.

Because of the recognized importance of these interactions, there have been a number of special observing and analysis programs focusing on specific mesoscale phenomena. From this preliminary work, it has been established that through synergistic effects, the interactions do indeed determine the spatial and temporal evolution of events on both scales. This initial research also indicates that because of the diversity and complexity of mesoscale phenomena, specialized numerical simulations that simultaneously resolve both the synoptic environment and the local weather are needed before the details of the interactions can be properly addressed. Such simulations will have to emphasize fundamental physics rather than parameterizations, and will have to be calibrated with specially collected data sets.

2. Some Questions

To illustrate the types of interconnections that exist between mesoscale and synoptic events, we give very brief descriptions of five different interactions. The first four of these examples are centered around the general theme of interaction between deep convection (as represented by cumulonimbus clouds or thunderstorms) and the synoptic scale, a theme that is only one of many that might have been selected. The fifth example addresses a more general theme, one that impacts on our ability to address the others. To emphasize that these are active areas of research, each example is introduced by a leading question of the type that might be posed by an investigator.

a. What physical process organizes the various cloud bands observed in hurricanes?

There is great interest in hurricanes both because of their potential for destruction and flooding of coastal regions, and because these tropical vortices play an important role in transferring thermal energy from near the sea surface to high in the troposphere. Observations from aircraft, from satellites, and with radar have revealed that a hurricane is a highly organized system of deep convection. The mesoscale structure consists of a series of inward spiraling bands of thunderstorms. These bands have the appearance of instabilities encountered in rotating boundary layers, but the fundamental cause remains obscure. The main spiral pattern is of particular interest because of the intense rainfall associated with the constituent thunderstorms. An understanding of this structure may also provide clues toward the mechanisms determining the motion of the parent tropical system.

    b. What is the coupling between a thunderstorm and the wind field of the environment that determines the direction and speed of storm motion?

Thunderstorms are routinely observed that move to the right (and in a few cases, to the left) of the mean tropospheric wind. Such deviation with respect to the mean wind has been found to be an indicator of a very strong thunderstorm, one with high probability of producing severe weather at surface. It appears that the updraft region of such a storm is in rotation and is of such intensity that it acts as a barrier to the horizontal flow. While both the development of rotation and the deviation of storm motion with respect to the mean wind have been related to the turning of the tropospheric winds with height, there is as yet no comprehensive theory for thunderstorm motion.

    c. What are the relationships between the fluxes of heat and moisture observed near the surface of the Earth and the evolution of mesoscale systems?

Such relationships have been extensively investigated in the case of the west Texas "dryline". This linear mesoscale feature is the narrow but very well-defined boundary between the hot dry air from the southwestern deserts and the warm, very moist air flowing inland from the Gulf of Mexico. Its interaction with synoptic-scale cold fronts moving southward frequently triggers explosive deep convection. Investigations of the dryline clearly show the controlling influence of surface temperature and moisture on the diurnal movement of this feature. The extension of findings concerning the dryline to parallel phenomena in other regions of the world has yet to be made. More generally, these findings suggest that surface fluxes may play a strong role in other mesoscale phenomena.

    d. What are the feedback mechanisms acting to transfer momentum, heat, and moisture from a Mesoscale Convective Complex to the corresponding mean synoptic fields?

These large systems of thunderstorms are an example of deep convection loosely organized on area basis. A mid-continental phenomenon, they have been observed to occur from mid-spring to late fall but are most common in late summer and early fall. (Thunderstorms organized on a linear basis as a "squall line" also occur throughout the same eight-month period, but are most common in early spring to mid-summer.) Mesoscale Convective Complexes have a life-cycle of many hours and typically travel as an organized entity many hundreds of kilometers. Because they persist so long, convection transports significant moisture and heat high into the troposphere. In addition, momentum is also transported aloft as the environmental wind field is disturbed on a large scale. Both the mechanisms organizing these complexes and those by which moisture, heat, and momentum are transferred from the individual convective cells to the synoptic scale remain speculative and are poorly handled by most existing numerical models.

    e. What degree of resolution is necessary in numerical simulations of mesoscale phenomena forced by instabilities in the synoptic flow?

Since simulations of such phenomena must be initialized with observational data, careful specification of resolution requirements is essential in planning both research and operational measurement programs. The amount of required data appears to be enormous and will come from diverse sources at high rates. This poses interesting problems in the development of techniques for efficiently producing merged or composite data sets. The data are likely to be "noisy", so that these techniques must incorporate automatic error checking and quality control procedures. The presence of noise also prompts consideration of its impact on the simulations and the degree of redundancy required in the data to compensate.

3. An Opportunity and a Challenge

We close by expanding on the fifth question raised above. As was noted in the introductory material, up to this time it has not been practical to routinely collect, using conventional instrumentation, the data needed to address questions like those posed in previous sections. This situation is changing rapidly as the field of meteorology is entering an observational revolution, one that promises to provide for the first time routine, high-resolution data on mesoscale events over a large area. A characteristic of the observation technology now being fielded is the production of streams of data that describe spatial or temporal distributions rather than isolated point values. As examples, the U.S. National Weather Service is beginning the deployment of the following four observing systems:

Automated Surface Observing System (ASOS). The automatic weather stations to be deployed as part of this system will provide around the clock observations from surface locations without the need for a human observer. The data will flow automatically to central collection points.

Next Generation Radar (NEXRAD). In addition to providing measurements of the reflectivity of rain within clouds, these new radars will also utilize Doppler techniques to measure the component of raindrop velocity lying along the line of sight to the radar. The sensitivity of the receiver is such that velocity measurements can also be accomplished in the clear air, using small density fluctuations as scatterers. Over 100 of these continuously scanning, surveillance-type radars will be deployed in the continental United States to monitor the development of all forms of weather.

PROFILER. This sensor is composed of a vertically pointing millimeter-wavelength radar and a very sensitive radiometric receiving system. Using phased-array techniques to steer the beam slightly away from the vertical, the radar can provide a vertical profile of the three components of the wind through the full depth of the troposphere every five minutes. The radiometer determines the vertical temperature profile at the same time by sensing the emissions from water molecules. (The current synoptic network uses balloon-borne systems to measure vertical profiles of horizontal wind and temperature once every twelve hours.) Coupled with the horizontal information produced by the NEXRAD system, the data collected by the PROFILER network promise to reveal the full four-dimensional structure (three space dimensions plus time) of flow fields associated with many mesoscale phenomena.

Advanced Geosynchronous Satellite (GOES). The next generation of geosynchronous satellites will incorporate a suite of sensors with new capabilities and improved resolution. Because the satellites will be three-axis stabilized, the sensors will stare continuously at the Earth and so give more images at improved resolutions. A particularly important sensor is a microwave radiometer or "sounder" that detects the amount of water vapor present in an atmospheric column. Of interest in its own right, the horizontal field of water vapor serves to map out many features on both the synoptic scale and the mesoscale.

The data from these new observing systems hold the promise of allowing the researcher to see the interactions between the synoptic scale and local weather events in all their complexity. However, to utilize these data in a research mode for addressing questions concerning scale interactions will require that the separate streams be merged (in real time) into composite data sets, archived, and ultimately analyzed to reveal the underlying physics. The efficient handling and productive utilization of mesoscale data in this quantity is perhaps the greatest challenge facing the meteorological research community for the rest of this century.

OCEANOGRAPHY AND GEOPHYSICS

D. Lal*
Scripps Institution of Oceanography
University of California, San Diego

Preamble

In Earth sciences, as in several other disciplines, we are currently witnessing a revolution. The central theme of the revolution seems to be an intellectual urge to synthesize data in the framework of consistent coupled large system models for processes occurring in the atmosphere, oceans and the lithosphere on different time scales. The process of the grand synthesis, like that achieved by the concept of plate tectonics, is now ongoing in several areas of geophysics and oceanography.

It is generally difficult to predict the direction in which a science is going, and then to make a prediction when one is in the middle of a revolution is certainly a very hard task. In Earth sciences, one can clearly see the trends and recognize the important problems of today, but it would be erroneous to presume that these would remain the key issues over a decade. They change dramatically as new concepts emerge. The time scales for major concepts to emerge has been considerably reduced in recent decades particularly because of the availability of high precision global data, and the easier availability of high speed number crunchers. We are now in a position to quickly check on global scale physical, chemical and biological models. Models have predictability and this is an important aspect of academic studies in geophysics; they are most valuable in studying the global biosphere-geosphere change.

The task of predicting where our science is going is made even harder because of an insurmountable obstacle, the personal equation. Scientists have their own personal prejudices -- everyone believes that the most important science of the future is what he is engaged in. I attended two international meetings seven years ago: one was on the history of oceanography and the other on the future of oceanography. The two meetings were held in tandem; both had a high level. The emergence of ideas in the past were very well covered, but when it came to the future, one could not help seeing a semblance between their projections and their own recent scientific work. Not unexpected, because if it was not the case, one would not be working on the problems at hand.

---

* Also at Physical Research Laboratory, Ahmedabad, 380009, India

The task assigned to me was: "What are the current most important questions in oceanography and marine geophysics?" What I present here is based on discussions with a number of active oceanographers and geophysicists. I do believe that key problems have been identified in these areas, and suitable actions are being taken to find answers to them. The Quo Vadimus symposium has been formulated a la Hilbert, who in his celebrated lecture before the International Congress of Mathematicians in Paris in 1900, listed 23 problems which seemed to him to be of fundamental interest. Hilbert then remarked "Who of us would not be glad to lift the veil behind which the future lies hidden; to cast a glance at the next advance of science and the secrets of its development during future centuries". Mathematicians now recognize that Hilbert did identify key problems of the future. In turn, we can call today's mathematics a mature field since most of Hilbert's problems have answers. Today the wisdom of scientific community is at an adequate level to undertake a Hilbert type challenge in the fields of oceanography and geophysics, but it must be realized that the problems here differ vastly in their character than in physics or in mathematics. The terrestrial phenomena represent time evolution of interactions between radiation and matter on a wide range of space and time scales. Understanding these problems replete with non-linear processes is a horrendously complex task. The future Earth can never be predicted but any given past record can be reasonably well understood in terms of key processes. Thus studies in Earth sciences largely aim at understanding present and past global scale phenomena and identifying principal processes and controls. The closest we can come to Hilbert would be to pinpoint some of the key elements in the system.

In the following we first present a broad brush view of the present outlook in the areas of oceanography and geophysics and then list some key problems which could be called the principal goals of studies in the next 2-3 decades.

Newly Developing Frontiers in Oceanography and Geophysics

In physical oceanography, efforts are being made to increase the coverage and accuracy of measurements so that temporal changes can be documented and understood. The World Ocean Circulation Experiment (WOCE) is planned with this idea. The satellite altimeter, already established as an instrument for blanket coverage and high precision observations of sea surface, will allow measurement of the surface pressure distribution and therefore the geostrophic current. The altimeter along with the scatterometer and the infrared radiometer will provide data to determine the large-scale ocean circulation, which is a super-problem containing a large number of important sub-problems, e.g., heat transport to polar latitudes, sources of eddy energy in the ocean, water conversion rates and processes.

Another important problem in physical oceanography is that of fluctuations. Non-linear systems, e.g., fluid flows in general often have the characteristics of exhibiting chaotic behavior. The complicated atmosphere-ocean system seems to exhibit rather long period fluctuations, without any external driving forces of such long periods; El Niño is an example of this. An adequate description of the oceans is necessary to identify the fluctuations. Prediction is especially important for management of man-made (and natural) chemical changes in the atmosphere, and their effects on climate.

Most of the geostrophic kinetic energy of the oceans is contained in the mesoscale (a few hundred kilometers) and time scales of the order of a few months. Recent developments in acoustic tomography seem therefore to be very vital for studying changes in the oceanic state.

In the field of chemical and biological oceanography, large-scale satellite monitoring of productivity (based on chlorophyll-related light bands) and the use of natural and artificial tracers to study air-sea exchange, changes in the chemical composition of sea water, and the principal biogeochemical cycles seem to be the new promising areas of study. The geochemical cycles of most elements in the oceans are strongly influenced by biological activity. This in turn is related to nutrient cycles, which depend on the large-scale ocean circulation. Climatic changes provide strong positive and negative feedbacks to the oceanic chemical and biological processes, as is evident from the oceanic paleorecord. It is, however, not clear how the changes are brought about; more specifically the cause and the effect are not well understood today. The present day oceanographer has a tough problem at hand. He has to understand the chemical and biological paleorecord, in conjunction with the observations of processes which are now taking place in the oceans as a paradigm, and climate as the cause as well as the effect of changes in the physical, chemical and biological state of the ocean.

In the field of geophysics, the structure and the composition of the ocean crust, the continental shelves and the continental basins have attracted the most attention. High sensitivity seismic refraction and multichannel seismic reflection measurements, singly or in combination have proved very rewarding. These have provided new insights into the formation and evolution of oceanic lithosphere by revealing, in a few places, magma chambers under oceanic spreading centers, and by documenting major crustal thinning under fracture zones in the central Atlantic. The bulk of the Earth's crust and lithosphere is formed at volcanic mid-ocean ridges. It is therefore very important to map the extent of the crustal magma bodies underlying the rises, and thereby to understand the evolution of the lithosphere as it spreads and ages. Recent OBS microseismicity studies at slow and fast spreading ridges have suggested the exciting possibility that the primary structural difference between slow and fast spreading ridges is the thickness of strong, seismically active lithosphere across the spreading axis. Recent theoretical studies of the causes of median valley topography suggest that the presence of strong lithosphere across the spreading center may be the structural requirement for its presence or absence. Coupled OBS microseismic and active source electromagnetic observations should allow one to test this hypothesis.

Studies of the extensive wave trains associated with the long-range propagation of the oceanic $P_n S_n$ and T phases have revealed a lithosphere as thick as 200 km in the northwest Pacific. Seafloor instrumentation capable of operating over increasing bandwidths for long periods of time are becoming increasingly important in providing the data needed to develop an integrated view of the structure and dynamics of the planet.

Whereas the dynamic nature of mid-ocean ridges is well established, no efforts have yet been made to document rates of change in the system components, nor the interactions linking the physical, chemical and biological processes involved. The complex and interrelated magmatic, deformational, hydrothermal and biological processes of operating at ridge crests span a broad range of time and space scales. A wide variety of coordinated and synchronized measurements will have to be made to permit understanding the system as a whole.

So far, plate tectonics provided a kinematic description of what happens to the surface of the Earth. Now we have to understand the fundamental causes of these surficial plate motions. Our thinking and development of instruments is going in the correct direction to achieve this task. The structure in mantle convection is defined by gravity, geoid, bathymetry, seismic, electrical and heat flow studies. Seasat data have shown that small-scale (200 km wavelength) convection occurs below the oceanic plates. Geoid measurements provide a natural complement to seismic tomographic measurements for mapping the density structure within the mantle. Such data should allow deducing the flow patterns of deep mantle convection.

Synthesis and Quo Vadimus

A number of international working groups in climate oceanography and geophysics are currently addressing themselves to the important observations which must be made to allow quantitative modelling of global processes. Two experiments (i) TOGA, which seeks to study the interannual variability of the Tropical Oceans and the Global Atmosphere, and (ii) WOCE, the World Ocean Circulation Experiment, have been proposed jointly by the Joint Scientific Committee (JSC) and the Committee on Climate Changes and the Ocean (CCCO). The Inter-Union Commission on the Lithosphere (ICL) aims at study of the dynamics and evolution of the lithosphere and is promoting experiments on a global scale. A number of workshops and working groups have concentrated on identifying important problems which must be studied to improve our current understanding in these areas. Besides the groups mentioned above, the deliberations of two workshops very relevant to our discussions:

(1) COSOD-II workshop on Global environmental changes: scientific goals of a drilling program designed to understand the mechanisms of change, and

(2) Workshop on the mid-oceanic ridge -- a dynamic global system.

References to the reports published by the various

organizations are given in this paper. We would, as examples list here some of the goals and objectives defined by WOCE and these two groups.

The WOCE Science Steering Group has set two goals for the experiment.

Goal 1: To develop models useful for predicting climate change and to collect the data necessary to test them.

Goal 2: To determine the representatives of the specific WOCE data sets for the long-term behavior of the ocean, and to find methods for determining long-term changes in the ocean circulation.

Within these goals, there are several specific objectives such as:

Under Goal 1:

(1) The large-scale fluxes of heat and fresh water, their divergences over 5 years, and their annual and interannual variability.

(2) The dynamical balance of the World Ocean circulation and its response to changing surface fluxes.

Under Goal 2:

(1) To identify those oceanographic parameters, indices and fields that are essential for continuing measurements in a climate observing system on decadal time scales.

(2) To develop cost effective techniques suitable for deployment in an ongoing climate observing program.

The unifying goal of the RIDGE (Ridge Interdisciplinary Global Experiment) initiative is to understand the physical, chemical, and biological causes and consequences of energy transfer within the global ridge system through time and space. This very broad goal has six specific objectives.

(1) To understand the flow of the mantle, the generation of melt, and the transport of magmas beneath mid-ocean ridges.

(2) To understand the processes that transform magma into ocean crust.

(3) To understand the processes that control the segmentation and episodicity of lithospheric accretion.

(4) To understand the physical, chemical, and biological processes involved in the interactions between circulating seawater and the lithosphere.

(5) To determine the interactions of organisms with physical and chemical environments at mid-ocean ridges.

(6) To determine the distribution and intensity of mid-ocean hydrothermal venting and the interaction of venting with the ocean environment.

The discussions above give a good feeling for the complexity of the problem which must be addressed in understanding how the global atmosphere-ocean-lithosphere system works; a large number of vested problems. For the present discussions, I pose two questions about the dynamic system of outer layers of the Earth:

(i) What processes control the climate of the Earth?

(ii) How does the environment of the Earth control the large-scale transport and mixing of materials in its outer layers?

These are in fact the two Hilbert type questions. These are no doubt mega problems but they are tractable in the coming decades. I would like to deal specifically with the climate problem only since sufficient progress has been made in this area.

One of the well established facts of long-term climate change today, is the dominant influence of the Earth's orbital variations on the pleistocene climate. Recently it has been shown that this is also true for non-glacial epochs. This is based on time-series analyses of several climatic parameters in marine sediments, (i) carbonate content, (ii) aeolian material accumulation rate, and (iii) oxygen isotope ratios. This raises several important climate questions because insolation changes due to changes in orbital parameters are small indeed. What are the non-linear interactions between climatic feedback mechanisms? How does small orbital forcing produce large amplitude climatic oscillations? What are the mechanisms by which small changes in the geometry of the Earth's orbit cause large changes in the volume of ice, the temperature and circulation patterns of the atmosphere and ocean, and the carbon dioxide content of the atmosphere?

In the coming decade, an entry to these important questions will come from several directions, e.g.:

(i) interannual variability studies

(ii) studies conducted under WOCE (fluxes of heat and fresh water; water mass conversion, etc.)

It may appear that these difficult problems may not have a plausible solution even after 3-4 decades. However we have reasons to be optimistic because we have at hand a large record of the past history of the Earth, which is now being read in greater detail with the development of new physical and analytical methods. As an example Broecker (1987) has proposed that rapid jumps in climate which have occurred in the past, e.g. the evidence based on ice record of oxygen isotopes and CO_2 content of the atmosphere, may be a result of a sudden dramatic change in the sea-to-air heat transfer conveyor belt mechanism. The marine sediment record seems to support this postulate, as evidenced by the great reduction in the formation of North Atlantic Deep water during glacial times. If further analyses support such a hypothesis we would learn at least about one of the ways in which abrupt climatic changes can be induced. The climate is apparently the net result of several transfers of fluxes and energy on a global scale, and is poised delicately to come to a new state of equilibrium with small forcings.

In all aspects of oceanography and geophysics, climatic changes, be they the cause or the effect, will be principal focus of studies in the coming decade. The international global biosphere change program of ICSU is a testimony to both the realization of existence of strong links between many branches of geophysics and oceanography, and the effects of this coupling on the global climate. There seems now the possibility that realistic coupled atmosphere-ocean-continent models can be developed in the coming decades and that suitable high precision global scale data would also be available to test these models.

It would indeed be a great achievement of science if the outer layers of the Earth can be conceived of and parameterized as a single dynamic system in the coming decades.

Acknowledgements. I am grateful to several scientists for their contributions to the problems discussed in this article. In particular, I would like to thank G. Arrhenius, W. Berger, C. Cox, J. Imbrie, J.D. Macdougall, W. Munk and J. Orcutt. Special thanks are due to Dr. John Imbrie and John Orcutt for making available texts of documents of their working groups and also for their substantial contributions to the present discussions.

References

(i) Papers.

Woods, J.D. 1985. The World Ocean Circulation Experiment. Nature, 314, 501-511.

Mason, B.J. 1987. The World Climate Research Programme. Contemp. Phys., 28, 49-57.

Broecker, W.S. 1987. The biggest chill. Natural History, 10/87, 74-82.

Orcutt, J.A. 1987. Structure of the Earth: Oceanic crust and uppermost mantle. Revs. of Geophysics, 25, 1177-1196 (U.S. National Report to IUGG 1983-1986).

Anderson, D.L. 1984. The Earth as a planet: paradigms and paradoxes. Science, 223, 347-355.

Flinn, E.A. 1982. The International Lithosphere Program. Eos, 63, 209-210.

Maxwell, J.C. 1984. What is the lithosphere? Eos, 65, 321, 324, 325.

(ii) Reports.

Report on the Ridge Program. Ocean Studies Board, National Academy of Sciences Publication. Washington, DC 20418. 1988 (in press).

"Whither the Oceanic Geosciences" a report by the Commission for Marine Geology. 1983. International Workshop on Marine Geosciences, Heidelberg, F.R. Germany, July 19-24, 1982. ed. J. Thiede, Oslo, Norway.

"Transient Tracers in the Ocean". W.M. Smethie, Jr. (editor), 1985. J. Geophys. Res., 90, C4 and C5, (AGU Publn).

Scientific Plan for the World Ocean Circulation Experiment 1986. World Climate Research Program Publn. Series 6. WMO/TD no. 122, WMO Geneva.

"The Global Climate System Autumn 1984-Spring 1986". World Climate Data Programme, WMO, Geneva. 1987.

"Earth Observing System" Science and Mission Requirements Working Group Report Volume 1. NASA Tech. Memorandum 86129. Goddard Space Flight Center, Maryland.

International Lithosphere Program. Newsletters 1-7. 1981 to 1986. ICSU Inter-Union Commission on the Lithosphere Secretariat Publn.

"A comprehensive program for solid Earth science Program". Earth System Science Committee, Geophysics Panel, NASA Advisory Council. 1986. Office of Space Science and Applications. Washington, DC 20546.

"Global Change in the Geosphere-Biosphere" -Initial Priorities for an IGBP National Research Council. National Academy Press, Washington, DC 1986.

EARTH SCIENCE IN THE 21ST CENTURY

Paul H. LeBlond
Department of Oceanography
University of British Columbia

Where is Earth science going? An overview of current trends clearly suggests some directions in which the Earth sciences are evolving today. Where these trends will lead us in the next century is the subject of this essay. Expected changes in the role and status of Earth scientists are discussed, as well as some of the problems to be encountered and the conditions under which these changes will take place.

The Evolution of Earth Science

Two overwhelming tendencies dominate the current evolution of Earth science: internationalization and unification. The first arises from the obvious recognition that natural phenomena do not acknowledge political boundaries. It has led scientists from every branch of Earth sciences to gather at first in *ad hoc* groups for carrying out specific programs, and later to institutionalized international cooperation by creating formal structures for collecting, archiving, sharing and interpreting information on Planet Earth.

Gradual unification of the many disciplines devoted to the study of various terrestrial sub-systems has been forced by the discovery of important linkages between them. This process includes three stages. First, the physical-chemical scientific disciplines of geophysics: seismology, vulcanology, aeronomy, meteorology, oceanography, etc. . . have been drawn closer together in joint projects, meetings and societies. Studies of other planets more easily perceived from a holistic point of view, have provided a strong stimulus for the unification of geophysics. A second fundamental step has been the recognition of the crucial importance of life processes in the development and maintenance of the Earth's environment. The International Geosphere Biosphere Program has consecrated the link between biology and the physical sciences in the study of the Earth. The third step is of course the inclusion of human activities as an important factor in the contemporary evolution of the environment; it has already opened the door to profound questions relating Earth sciences with economics, forestry, agriculture...all human activities.

In relation to the above structural trends, Earth sciences is maturing rapidly in terms of its understanding of the various sub-systems which contribute to the terrestrial environment, the interactions between them and the ability to foresee their future behaviour.

Earth Science Tomorrow

Looking forward half-a-century in the future, after the successful completion of the ambitious programs of climate prediction now underway, and after five more decades of structural evolution following the tendencies described above, one may readily imagine Earth science as a radically altered activity, in terms of its capability, role and status.

Earth science in the next century will be a giant compared to that of today, as is today's with respect to that of the last century. Its predictive power will be greatly increased. It is reasonable to expect, within a century, reliable prediction of atmospheric and oceanic climate on time scales reaching a decade. That is the goal of some of the largest cooperative programs now underway and it would be very pessimistic to think that success in that direction in the next few decades would not match that made in weather prediction over the past decades. One would expect significant advances in other fields, such as for example in warning against impending volcanic and seismic catastrophes.

Reliable prediction is of course based on adequate scientific understanding. It will also rely on a vastly expanded data acquisition system, in which satellites and other remote sensing devices will play a major role, communication systems and computer models. Individual and cooperative research will continue to improve the system. In the context of the unified and internationalized Earth science of the next century, it makes sense to refer to the core of the prediction system as the Earth Model. The Earth Model will have at its core an ocean-atmosphere evolution model, linked with a periphery of sub-systems and special sub-models. The Model will be redundant, distributed and interactive, continuously updated and refreshed with data inputs. It will grow naturally from the integration of current modelling and prediction practices and will become the framework upon which new theories are tested and grafted. The improvement, testing and challenging of the Earth Model will become a preoccupation of a growing number of Earth scientists.

Success brings prestige and influence. The hundreds of thousands of scientists and technicians involved in the Earth science of the next century may see their status enhanced. The organizations which speak for them will certainly play a strong role in Earth affairs. As Earth science expands into Earth Care and human activities play an ever more crucial role in the environmental balance, we may expect global environmental issues to rise towards the top of the political agenda and Earth scientists to play a stronger advisory and perhaps regulatory role. One may also envisage the creation of an International Earth Protection Agency empowered by international consent to review projects likely to have a significant impact on the terrestrial environment.

The Path to Tomorrow

The above scenario depicts Earth scientists as the wardens of the Earth, working in a spirit of international harmony to protect humanity from local catastrophes and to maintain its global life support system. How realistic is this rosy scenario? A future view based on extrapolation of current trends inevitably glosses over obstacles and difficulties.

The first and most fundamental milestone is the achievement of a high credibility level in the forecasting of natural phenomena of public interest: beyond tides and weather to seasonal and interannual climate conditions as well as prediction of earthquakes and volcanic eruptions. It is not sufficient that predictability be attained; it is essential that these predictions be believed and that the trustworthiness of Earth science be firmly established. The progress of Earth science depends as much on the perception of it by the public as on the achievement of its own goals. International unanimity is a crucial element in establishing the required level of credibility.

The status of Earth scientists in the next century, and the role which they will be called on to play will result in great part from the initiatives taken in the intervening years by individuals and their associations. Continuing and conscious efforts are needed to deepen the internationalization of Earth science and to develop the dialogue between Earth scientists and other groups with global concerns, in international law, pollution and the environment, for example. The fostering of a global perspective in Earth science curricula is to be encouraged, covering both scientific and historical points of view. Earth scientists must take the lead today in speaking for the protection of the planet if they are to be respected and followed tomorrow; their responsibility goes beyond individual dedication to collective involvement through their professional societies, scientific unions and other groups which they may wish to join or create.

Finally, a plethora of political, economic and cultural obstacles stands in the way of the peaceful world in which Earth scientists may stand as overseers of the planetary machinery. War is unkind to the environment; nuclear war would have global consequences. Economic motives too often leave environmental implications out of the balance sheet. Age-old practices can lead to catastrophic environmental effects when expanded to modern scales. Earth scientists need not remain passive observers of world events, nor view them as beyond their power and expertise. They already command sufficient respect to lead public opinion towards environmental awareness and conflict resolution. The implementation of the Atmospheric Test Ban Treaty is an example of the influence which scientific and public opinion can have on political issues. It is strongly in the interest of Earth scientists to take a leading role in initiatives which aim at creating political conditions under which their role and status would be enhanced. Again, not just individuals, but scientific and professional associations must be involved for maximum impact. World peace is clearly the optimal state of the human component of a planet in which Earth science may develop its full potential.

Conclusion

Under peaceful conditions, Earth science should naturally develop into a system for planetary stewardship sometime in the next century. It is largely up to us, Earth scientists, to make this happen.

PHYSICAL OCEANOGRAPHY TO THE END OF THE TWENTIETH CENTURY

R. W. Stewart

Physics and Astronomy Department, University of Victoria, Victoria, B.C. Canada V8W 3P6

Decadal predictions of the evolution of the science are likely to be as unreliable as similar predictions of the evolution of the climate system. In each case, the fundamental requirement for chaos is present: small causes do not necessarily lead to small effects; and paths in phase space, which are at one time close neighbours, may diverge very widely as time evolves. The uncertainty in the evolution of the science arises from the flash of inspiration to an individual scientist or from the totally unexpected result of an observation.

Nevertheless, in each of these systems inertia and the presence of known influences provide enough constraints that some level of prediction is possible. There will be no full-blown ice-age in this century. There is simply not enough time for the required mass of ice to build up, regardless of what happens to global temperatures. On the other hand, the accumulation of radiatively-active gases in the atmosphere will continue, hardly (if at all) abated. International socio-economics has too much inertia for the dramatic changes which would be necessary to arrest it in that time frame.

Similarly, there are constraints which limit the possibilities in oceanography for the rest of this decade. One of the most important of these constraints lies in instrumentation. I will say with confidence that no instrument will be in widespread use during this century which is wholly unknown to, or unimagined by, present-day oceanographers. This assertion is based on experience, and arises from two principal considerations:

1. For instrumentation where really large amounts of money are available for development and testing – notably instrumentation to be flown in satellites – the approval processes are so complex and the "queue" is so long that we already know the maximum to be expected in this century. We may lose some things, but we won't gain anything radically new.
2. With respect to other instrumentation – the kind that oceanographers invent, develop and produce either in their own laboratories or in conjunction with some (usually small) private company – the reasons are different but the result is similar. The time for development and testing, added to the time required to persuade the oceanographic community that they ought to adopt such an instrument, and then the time required to get large numbers produced, means that typically at least a decade elapses be tween the conception of an idea and its widespread adoption.

Another important factor is computer development. Here we have a double predictability: nothing really unheard of at the present time will be in full use within the century, for similar reasons to those offered above. However, what we have heard of is impressive. We may look forward with some confidence to continued growth in the computer power available to oceanographers. Front-line computers can be expected to continue more or less on the capacity growth-line experienced for the last several decades. The market is clearly there, so there is plenty of incentive for manufacturers. There seem to be no insuperable technical obstacles: at least one more generation of reduction in cell size for microelectronics seems likely, and the possibilities of multi-parallelism have barely been touched. Also, there are much more radical proposals being explored, some of which will undoubtedly prove fruitful.

Perhaps equally important for oceanographers is the "filling in behind" being undertaken by some computer manufacturers. Capability comparable with that of what are presently referred to as supercomputers is becoming available at much lower prices and can be expected to become much more widespread.

With respect to data storage and exchange, we seem to be on the verge of a major revolution with the development of laser-disk ROMs which will make less-processed data available to a much larger number of groups than is presently the case.

The other fact which gives us some predictability is the existence or planning of large-scale international programs. TOGA is underway. WOCE is in an advanced state of planning, and I am confident that it will happen. So will JGOFS and some other aspects of ICSU's Global Change Programme. So will the next stage of deep-sea drilling.

I will not spread my net so wide, but will confine my speculations to large-scale physical oceanography.

TOGA

Although the study of interannual variability of the Tropical Oceans and the Global Atmosphere (TOGA) was officially proclaimed to have commenced on January 1st, 1985, the build-up period started well before and continued well after. This build-up has been sufficiently gradual that it has not been widely recognized that what we are seeing in TOGA is a change in the face of oceanography, which will undoubtedly have major ramifications on the way in which the science is conducted in the future.

In TOGA, we now have a widespread system of observations, very little of which involves traditional methods of doing oceanography. Traditionally, oceanographers have gone to sea in specialized vessels, made observations with specialized instrumentation that only they and a select group of well-trained technicians are able to maintain, returned to home laboratories to work up their data, and reported the results in refereed journals and meetings and seminars. Instead, data are being collected with rather simple instrumentation which can be dealt with by people having only limited training. In other cases, data are collected from unmanned buoys and stations.

Data are flowing in and being made available to research workers within weeks – in some cases hours – of being collected. Routine analyses of certain kinds of data are being prepared and disseminated. Predictive ocean models, and to a limited extent coupled oceano-atmosphere models, are being run using these data.

Models run in a "hindcast" mode show features remarkably similar to those observed. Forecast models are being run and the more bold among those running them are making predictions of the future evolution of the tropical ocean and further predictions about the behaviour of the atmosphere under the predicted oceanic conditions.

Thus, for the first time, oceanography is a genuinely operational mode, albeit an experimental one; we still have the close involvement of scientists who have no intention of making a life's work of it. So far all of this is almost completely confined to the Pacific. There is a reasonable observational net in the Atlantic, but no body of theory comparable to that which exists in the Pacific has yet been developed. Indeed, the tropical Atlantic is so narrow that the two situations may not be very analogous.

In the Indian Ocean it has proved much more difficult to build up the observational net. Despite the shining examples being given by Australia and Mauritius, for some reason the riparian nations around the Indian Ocean have been unable to unwilling to take the steps necessary to put a comprehensive observing system in place. Problems in the Indian Ocean have been further aggravated by a chronic inadequacy of data derived from geostationary satellite coverage over that ocean, despite the presence of INDSAT

The meteorological effects of major shifts in oceanic regime of the Pacific Ocean are so dramatic and so widespread that the phenomenon of "El Nino" has caught the popular imagination. From a scientific point of view, the study of that phenomenon is in that most productive of all phases: theoreticians are using recently-generated data and are modifying their theories on the basis of these data. Those planning observations are aware of the theories and design the observing nets in their light.

It is my expectation that this aspect of the science will develop rapidly. Ten years from now there is a high probability that most of the theoretical notions now being worked with will be regarded as being naive and primitive, and many of the observational programs will be regarded as having been ill-judged and badly designed. But these future judgments will be made in hindsight on the basis of greatly improved understanding. By the end of the century, equatorial ocean models will probably be used routinely in making seasonal climate fluctuation forecasts, and I believe that these forecasts will show a skill at least comparable with extended-range weather forecasts at present. TOGA will probably not end in any formal way, but will gradually shift from an experimental into a fully operational mode, reminiscent of the present World Weather Watch.

It is hoped that the success of TOGA in the Pacific will at last convince nations with great interest in the Indian Ocean to apply a similar effort. Perhaps they will achieve a comparable result, although I do not really anticipate that they will soon overtake those working with the Pacific. Nevertheless, I am optimistic that the next decade will show a far better understanding of the nature of fluctuations in the monsoon than we now have, and that there will be some useful predictive capability, again based on an understanding of the behaviour of the ocean.

In the meanwhile, the World Ocean Circulation Experiment (WOCE) will get underway. If it starts as planned in 1990, that start will almost certainly be more nominal than real. Organizing these large international programs always taken much longer than is expected. (Remember that the Global Experiment of the Global Atmospheric Research Programme was initially (1967) envisaged as taking place in 1973. It was run in 1979.) I am, however, convinced that it will happen. The two major driving forces which led to initiation are unabated:

1. The inexorable increase in the concentration of radiatively-active gases in the atmosphere ensures a very high-level international concern about possible major climate changes. Knowledge is becoming increasingly widespread that no prediction of such a climate change will be possible without much more through understanding of the behaviour of the ocean than we now have. Media people and politicians are now aware of these facts.
2. The scientific drive to understand the working of the

whole ocean, so that one may consider the effects and the importance of the mulitiplicity of processes which have occupied oceanographers for the last three decades is, if anything, growing. Too many people believe that we must try for this effort to dissipate. Also, it now seems very probable that we are going to get the satellites which were always declared to be sine qua non for this experiment. We will get the altimeter. Probably Topex-Poseidon, but if not, something not so different. We will get scatterometers. Both of these will be useful in TOGA and in looking at regional problems, but above all they will encourage a global look at the ocean. The altimeter will definitively resolve the data problem for deep-sea tides and for time-dependent surface geostrophic flow. The scatterometer will provide full coverage on a set of phenomena related to the wind driving of the ocean − although I anticipate that the translation of scatterometer cross-sections into wind stress will prove more complex than many people now assume.

By the mid 1990's, global coverage of data collected in the interior of the ocean will be substantially greater than it is now, and new data will be coming in apace. There will even be a few acoustic tomography nets in place, to obtain for the first time continuous data on the behaviour of the ocean interior.

The inverse techniques of data analysis, which have been so extensively developed in the last few years, will be used widely and will be considerably refined. These techniques, which permit the objective reduction of uncertainty in estimates of the values of important quantities, are particularly appropriate to a situation where new data are being obtained and there is the opportunity to modify data-gathering programs in such a way as to further reduce uncertainty in a more nearly optimal fashion.

Undoubtedly differences of opinion will arise over interpretation. These differences will provide added incentive for more work: observational, on data analysis and on theory. The increase in the availability of data and computer capacity, which was alluded to above, will mean that far more groups will be active in this work than is presently the case. New notions about the way in which the data should be interpreted, and new methods of data analysis will arise. We can expect a rapid advance in the science.

It is in this area that predictive glasses are most clouded. While it may take ten years to get a new instrument into widespread use, it may need only as many months for the dissemination of a new idea. It would be foolhardy indeed to attempt to predict the evolution of ideas, except to say that the accepted view of the behaviour of the ocean will surely be quite different in the year 2000 than it is now. We will know more, but more importantly, we will understand more. By the year 2000, we should be in a position to undertake a serious attack on predicting the nature of the coupled ocean-atmosphere climate system, with different concentrations of radiatively-active gases than we have at present. We should, by then, have ocean-wide models with grid resolution small enough to handle ocean eddies. We should have greatly improved physics in these models. Our knowledge of air-sea interaction should have improved somewhat, so that the uncertainly in parameterization of heat and momentum flux will be appreciably less than it is now. (I am not as sanguine about developments in this area as I am in some others. There are just not enough people working on it in the world, in comparison with the difficulties of the problems. I have some hope that the undoubted differences which will arise will scatterometer interpretations, because of the different wave lengths being used in European and American satellites, will lead to increase funding for and activity in this area.) By then, there should also be well-established modelling techniques for handling the difference in characteristic time scale of the ocean in comparison with the atmosphere (surely by being more clever than using the brute force technique of running the ocean model on atmospheric time scales − even though by then we may have enough computer capacity to make it possible to contemplate such calculations).

More and more modelers will have the temerity to cast coupled models adrift and to let them go where they will. I anticipate that these models will go in strange directions indeed, and generate climates quite unlike our present one. I also anticipate that there will be vigorous debate as to whether these climates represent something real in the sense of Lorentz' quasi-intransitivity, or whether they merely advertise inadequacies in the models. Therefore I expect that the models used to advise governments will continue to have a number of "tie-downs" to keep them from wandering too far from present reality.

Resolution of this problem will be one of the tasks facing the next generation of oceanographers, in the next century. Over a decade will have expired, and we may assume that they will have the advantage of new tools not now imagined, and a pace of data gathering far beyond anything we have experienced. They will also have a stock of historical data larger by orders of magnitude than anything we now possess and commensurate techniques for dealing with these huge volumes of data.

The 90's promise to be an era of great excitement in physical oceanography − with the science moving powerfully in the direction of becoming operational, predictive and global.

COMMENTS ON R.W. STEWART'S "PHYSICAL OCEANOGRAPHY TO THE END OF THE TWENTIETH CENTURY"

C. Wunsch

Massachusetts Institute of Technology, Cambridge

My comments on this note are mostly general. The major one is that in the context of a Hilbert-like discussion of the "key problems" the paper doesn't really do that at all. Rather, Stewart is mainly speculating on what the state of the subject is likely to be about the year 2000, along with some description of the trials people will have in doing it. As such, it is reminiscent of more elaborate efforts along the same line (I believe that SCOR produced an entire document, under K. Hasselmann's editorship, only four or five years ago, that talked about oceanography to the end of the century.)

Furthermore, all our experience in oceanography is that a decade is a very short time for major changes. It is exceedingly optimistic to use 10 years as the time for an instrument to go from conception to widespread use; I believe that the true timescale is closer to twice that (there's a long list of examples, including current meters, CTD's, etc.)

I really have no particular quarrel with Stewart's views however. But if one were to emulate Hilbert, I would have thought a discussion of significant hard problems to be solved in the next century would have included:

-- mixing with all its ramifications, including scale dependence, boundary enhancements, etc.
-- air-sea transfers, including sensible and latent heat; gas exchange, etc. (Stewart alludes to this.)
-- where and how the tides are dissipated.
-- detailed understanding of the interactions in the internal wave field, etc.

Everyone would have his own list.

One could also list technical problems:

-- determining evaporation minus precipitation from space.
-- the construction of unmanned observing platforms.
-- solution of the telemetry problem, etc.

QUO VADIMUS -- HYDROLOGY

Zbigniew W. Kundzewicz
Institute of Geophysics
Polish Academy of Sciences
Warsaw, Poland

Introduction

It is not a new idea to prepare a list of specific current problems of scientific hydrology. One could mention perhaps three endeavors of this kind ending with publications. The Committee on Status and Needs of Hydrology, convened by the American Geophysical Union, prepared a list of sixty-three challenging research areas as early as 1964 (Linsley, 1964). Although much research has been directed to these problems, most of them remain current because new pressures on hydrological systems require higher levels of sophistication and accuracy in problem solving. Examples of key problems identified by Linsley (1984) that are still in the forefront of attention of hydrologists are related to the basics of hydrological processes, such as flow in unsaturated porous media, evapotranspiration, effects of land use, and the synthesis of the hydrological cycle. Other problems ceased to be the most challenging research topics. Examples of areas which are not of great current interest include: problems of flood routing; unit hydrograph simulation; information theory applied to hydrology; or, to some extent, hydrological systems analysis. In the time period since the publication of the list by Linsley (1964), these areas have yielded some useful (and still used) procedures, but are not in the forefront of hydrological sciences anymore.

In 1986 a comprehensive special issue of Water Resources Research was published, devoted entirely to the analysis of trends and directions of hydrology. Fifteen papers in the volume, written with a philosophical and reflective flavor, offer a range of research needs that present a challenge to the researcher in hydrology (Burges, 1986).

Most recently, dozens of specific current research problems were discussed in the Report of the HYDROLOGY 2000 Working Group acting under the International Association of Hydrological Sciences. The latter report (Kundzewicz, Gottschalk and Webb, eds., 1987) was presented and discussed during the XIXth General Assembly of the IUGG in Vancouver (Hydrology 2000 Workshop, Aug. 17th 1987). Some ideas that will be mentioned herein are elaborated in more detail by the writer's colleagues from the IAHS Hydrology 2000 Working Group and by the writer himself in the Hydrology 2000 Report (Kunzdewicz et al., 1987).

It is felt, that the number of Hilbertian-like problems that have been identified in hydrology is high (hundreds, maybe thousands). Several specific current research problems were discussed in abstracts of other hydrology contributions to the Quo Vadimus Symposium. Examples of such problems are briefly presented herein. However, it is not the purpose of this contribution to provide a long list of particular research needs. Instead, a few, rather general, challenging issues will be mentioned that embrace many specific unsolved problems.

Hydrology is concerned with the study of water which focuses on the investigation of the processes contributing to the hydrological cycle. However, the scope of hydrology has been traditionally focussed on the terrestrial and atmospheric rather than the oceanic phases of water circulation in nature.

As hydrology is a basic science within the family of geosciences, its primary issues are cognitive; to improve the understanding of laws governing the phenomena and processes contributing to water circulation, understanding of spatial and temporal distribution of water resources and water quality aspects associated with the physical, chemical and biological components.

Besides being a field involving a large number of basic research areas, the study of hydrology serves utilitarian needs. The hydrologist contributes to alleviating increasingly complex problems connected with scarcity, abundance and pollution of water. Therefore, the concerns of hydrologists have always transcended the boundaries that would be appropriate if hydrology were to be confined only as a branch of geophysics (Dooge, 1984). The practical aspects of hydrological sciences reflect the importance of water, a major constituent of living matter and which is necessary, on a continuous basis, to sustain the processes of life and it is essential to virtually every human endeavor. Practical problems that hydrology is faced with result from demographical, technological, agricultural, social, political, economical, and aesthetical considerations. They primarily pertain to the hydrological aspects of the use and management of water resources.

Two basic utilitarian needs that the study of hydrology is bound to address include the assessment of water resources and the dynamics of their occurrence. This assessment includes quality aspects, spatial and temporal variability, and improved predictions of future conditions at several time horizons of interest.

Bearing in mind the above basic science and utilitarian aspects, one can understand the future of hydrology as the trade-off between possibly conflicting priorities to the science and to society. The former pertains to the rise of the basic understanding of the subject of hydrology. It addresses the intellectual quality in long-term scientific enquiries with no immediate applicability. The obligations to society result from the need to immediately address the burning issues of alleviating water related problems.

Scale Problems in Hydrology

There is an urgent need to extend the range of scale levels of hydrological analyses. The scale levels traditionally dominant in hydrology pertain to a small drainage basin and this results from the practical problems of water supply and hazard reduction. However, as pointed out by Dooge (1987), differences of the concepts regarding water and models of water behavior, depend greatly on the problem of scale. Water can be studied at different scales, for example, at three micro-scales (molecular cluster, continuum point and representative elementary volume), at three meso-scales (module, sub-catchment and small-catchment), and at three macro-scales (large-catchment, continental and planetary).

It is desirable to conceive the hydrological processes at all scales ranging from molecular to global and to strive towards hydrological theories at several scales, providing sound models of different complexity, applicable to different purposes. It is also desirable to attempt to connect laws of behavior which are operative at different scale levels.

Although the partial differential equations of mathematical physics governing hydrological processes on a continuum point scale are deemed known, rigorous aggregation to a larger scale is very difficult in practice, if at all possible. Such an aggregation performed via integrating accurate laws throughout the domain of interest, would mean taking full account of the spatial variability of parameters, and irregularities of form, and temporal changes of systems and processes. This does not seem realistic due to the high spatial variability of hydrological parameters (e.g. ground properties) and nonlinear phenomena involved. Therefore the model can be formulated on a larger scale, being compatible with physical principles, or else -- without direct reference to the lower scale laws.

Independent of the rigorous physically based mathematical models, there are also other approaches widely used in hydrology. There are conceptual models, created in a subjective way, usually on the meso-scale level (e.g. for a catchment or a river reach) that represent some most important physical properties of the prototype. There are also empirical models and "black box" system approaches that attempt to extract information from input-output data without penetrating into the physics of the process, i.e. without unveiling the cause-effect relationships.

The advances in computer and remote sensing technology have made it possible to approach the problems of global hydrology (cf. Eagleson, 1986). The advent of global hydrology poses the problems of refining the estimates of global world balance of water, heat, and materials. In the area of global water balance, studied in hydrology for many decades, there is still no consensus as to how to estimate the global components for the earth and for the individual continents. Global water balance analyses have often not taken account of changes of the mean sea level and ice caps. Icemelt from glaciers and increasing sea level means addition of large volumes of water to the cycle that is often considered closed (Meier, 1983).

Similarly, it seems challenging to assess the volume of very deep groundwater stored in the earth's mantle and also to estimate the dynamics of exchange of this "inactive" water and the rest of the hydrosphere (Dooge, 1983).

The hydrological impact of large-scale anthropogenic changes needs to be assessed. This embraces both side-effects of human activities (like increased combustion of fossil fuels) and macro-engineering water projects (e.g. river diversion, drainage of large swamps). The mechanisms of propagation of these latter effects to distant regions via atmospheric dynamics (tele-connections) need to be understood.

Examples of specific problems falling into the perspective of large-scale or global hydrology were posed by Eagleson (1986). Will the macro-engineering water projects (e.g. drainage of large swamps) reduce the local precipitation, and by how much? What locations will feel the effects of such a land surface change? Additional questions that could be formulated read:
- What is the destination of water evaporating from here?
- What is the source of the water falling here as precipitation?

Tracer experiments and general circulation models are needed to provide answers to these questions.

Strive Towards Interdisciplinarity

At the infancy stage of the development of hydrology, systems and processes were conceptually isolated from their environments. The further stage of development of hydrological research requires reconstruction of broken interconnections and feedbacks, that would allow us to extend the scale level of analyses. This calls for joint consideration of hydrological (and other) subprocesses, for integrating separate approaches to different parts of water circulation in the Earth environment. Eagleson (1986) pointed out the need for the geophysicists's view of the large hydrological cycle, including the oceanic branch. The hydrosphere should be perceived as coupled with the atmosphere, lithosphere, and biosphere in complex interconnections. In order for these complex couplings to be considered, analysis of one single global supersystem is required.

It is strongly felt that hydrology, similarly to other areas of geophysics, needs "tearing down disciplinary barriers" (title of paper by Roederer, 1985). The domain of hydrology in the traditional, narrow understanding, does not cover areas that must be taken into account when trying to understand, describe, and forecast the behavior of complex hydrological systems.

In order to stimulate interdisciplinary approaches the university curriculae would have to be re-evaluated. This could allow understanding of more areas than were in the traditional background of hydrologists. Working out a common language of problem formulation, understandable by scientists of different disciplines and sub-disciplines is necessary for improved communication.

Examples of topics, that clearly belong to interests of several disciplines and which can illustrate the needs of inter-disciplinary and multi-disciplinary approaches are sediment production and transport. This topic appeals to civil and agricultural engineers, foresters, and land use managers, hydrologists, geologists and geomorphologists, estuary and coastal engineers and ocean scientists. Another topic would include the interface between water quality and human health (e.g. the risk of drinking water, biological availability of trace substances, factors that control their distribution among liquid and particulate phases) and requires collaborative works of hydrologists, geologists, chemists, biologists, ecologists, toxicologists, specialists in hygiene, and medical scientists. Another immediate inter-disciplinary challenge is studying the relationship between the CO_2 increase and heat balance, glacier behavior, precipitation, evapotranspiration, soil moisture, runoff and sea level. Improved communication with other disciplines is important for assessment of the consequences of climate change to hydrology and water management. This is one of the main present interests of hydrologists. The influence of climate changes and climatic variability on the hydrological regime and water resources has been the topic of hydrological symposium HS3 at the XIXth IUGG General Assembly in Vancouver, 1987. At this symposium several required inputs from other disciplines were identified to address the issue of atmospheric CO_2 increases. Hydrology needs forecasts of the consequences of increases in CO_2 concentration on evapotranspiration, functioning of stomata, plant water use, and biomass production. On the other hand, hydrology could contribute to global circulation models, where input in the area of subgrid parameterization seems welcome.

The notion of interdisciplinarity in hydrology takes a different form than in other geosciences. There are few university courses which will provide hydrologists a broad perspective of present and future hydrology issues. The discipline of hydrology is run primarily by holders of university degrees in agricultural engineering, civil engineering, environmental studies, forestry, geography, geology, systems science etc. This causes the obstacle to a unified science of hydrology. On the other hand, however, this variety of backgrounds gives hydrology a special position with regard to interdisciplinarity. Representatives of other disciplines are already within hydrology, and drawing from imported concepts and methodologies is relatively easy.

Some Credibility Aspects

The existence of basic research and utilitarian facets of hydrology tends to give rise to an unwelcome large gap between the two categories of methods and models. The methods used in practical applications, rather than accounting for the nature of processes, are frequently based on the empirical curve-fitting philosophy (e.g. multiple regressions) without physical foundations.

It is natural to challenge the credibility (accuracy and reliability) of the methods, and the data, and interpretations which result from these methods. The attempt to improve credibility could start from critical re-evaluation of assumptions upon which hydrological methods are based. Several assumptions have been taken arbitrarily, for the sake of convenience, elegance and tradition, rather than to adequately represent reality.

Certainly, the adequacy of assumptions is to be considered for the given scale of the problem. Assumptions at different scale levels can be mutually discrepant. Isotropy may be assumed at one level of analysis, whereas anisotropy assumption in studies at another level need to be made (e.g. Dooge, 1983). Similarly, depending on the scale level, one can take either of alternative assumptions, linearity or nonlinearity, stationarity or nonstationarity etc. This is consistent with the engineer's point of view -- to neglect effects that are not important in the analysis in question and to consider essential aspects only.

It is dangerous, that the growth of model complexity in contemporary hydrology accompanies, first of all, the development in the computer technology rather than the rise of our understanding. Therefore growth of accuracy of results is far lower than the complexity increase and the weaknesses of methodologies do not vanish, but rather hide themselves in another form.

Where extrapolation is involved, the overall results are not credible and it may be difficult to assess the degree of their credibility. It does not seem possible at all to assess the reliability of extrapolations in the case of empirical or "black box" system models. One should be aware of the "unreliability of reliability estimates" (part of a title of paper by Klemes, 1979) and not pretend that hydrology can offer reliable information concerning recurrence times of hypothetical very rare events. The urgent questions posed by the practice are often related to very rare events, or events that have not occurred at all. Examples of such information needs are relevant to the design of very large dams and of cooling systems for nuclear power plants, where the safety aspects require consideration of events with recurrence intervals of 10,000 or more years. In such a case, as noted by Klemes (1986a), the notion of a 10,000-years-flood has "mostly qualitative and public relations significance" and could well be replaced by terms like dangerous flood, catastrophic flood, and so on.

In such a case, application of soft approaches (fuzzy logic and expert systems) may be necessary and the eventual rationalizations should be based on assumptions, hypotheses, tradition, experience and intuition, rather than on the hypocrisy of a "scientific procedure" yielding a design flood.

The hope for progress of flood frequency analysis dwells on using some physical information rather than mere extrapolation of a series of numbers based on the assumption of homogeneity and stationarity of processes and a subjective choice of the probability distribution function. However, it is quite frequent that the series of observations are extrapolated into the regions where other physical mechanisms occur usually triggered in a threshold mode for very high flows. Examples of such mechanisms (Klemes, 1986a) are -- sudden increase of flood generating area, diversion of river flow from one basin to another, or flooding large detention areas.

Another aspect of extrapolation is use of methods and models outside the domain of their legitimate use. As an example, the unit hydrograph theory for small humid catchments has been treated as a universal tool applicable to any catchment. The unit hydrograph theory has been applied to large and arid basins with only very few point rainfall observations. There are several methods that, once proven successful in certain conditions are increasingly used, and gradually abused, in the bandwagon mode for other conditions and other processes.

Review of Hydrology Contributions to the Quo Vadimus Symposium: Examples of Specific Problems

There are five contributions to the Quo Vadimus Symposium in which varying hydrology perspectives are represented. The extended abstracts of these contributions are available in the preprinted abstract collection.

In their abstract, Duckstein, Nachtnebel and Bogardi (1987) assessed the problems of long range risk analysis, uncertainty, and fuzziness. The specific problems concern the choice of the appropriate representation of elements of hydrological systems (are they random, uncertain or fuzzy?), and the assessment of membership function and its role as an indicator of the effects of imprecise environmental consequences. How can the trade-offs be made to manage the risk? -- reads the final question by Duckstein et al. (1987). The above set of questions illustrate the needs of interdisciplinary backgrounds. Hydrologically useful concepts from risk analysis and fuzzy sets theory have not belonged to the arsenal of means that most hydrologists can recollect from their curriculae.

The abstract by Krausneker (1987) contains a large number (eleven) of "non-technical problems" pertaining to the element water, to its vital formative capacity, to philosophy of hydrology, to its contribution to water resources management, and finally to hydro-sociology.

Lliboutry (1987) formulated four key questions pertaining to the rheology of metamophic rock ice. This branch of glaciology indeed offers a number of intellectual challenges. Lliboutry (1987) asked for reasons why rock ice formed during the last glacial episode had much smaller grain size and was much more anisotropic than rock ice formed during the Holocene. He posed also various general questions which dealt with rheological parameters of anisotropic ice. He also posed questions dealing with general constitutive laws for transient creep.

The current problems presented by Rango (1987) concerned the interdisciplinary issues and the cognitive and applied perspectives of hydrological sciences. "How can the behavior of organic and inorganic chemicals as they move through the root, vadose, and groundwater zone, be better understood in order to predict their occurrence and concentration?" -- is the key problem identified by Rango (1987) for the time scale of the near term (0-15 years) needs. The key medium term (5-20 years) problem is -- how can new technological advances (e.g. microprocessors and remote sensing) be used to improve hydrological modelling? Finally, the long-term (15-40 years) problem is -- "what effects do large scale hydrological processes have on food production, weather and climate, and world-wide water supply?"

The main problem raised by Ubertini and Natale (1987) pertained to the meso-scale perspective of hydrological systems. They pointed out the necessity of joint, integrated formulation of hydrological models and the design of hydrological monitoring systems.

It is clearly seen, that although many specific current problems were identified in the five extended abstracts mentioned, they collectively cover only a small area of the domain being of interest to hydrologists. This gives the feeling of the fragmentation of perspectives of individual hydrologists and of the broadness of the domain of hydrology as a whole.

Concluding Remarks

There is some evidence that classical hydrology has reached barriers in its progress. The way out is seen in the extension of scale levels of hydrological analyses up to the global scale, and connecting laws of behavior operative at different scale levels.

It is strongly felt, that interdisciplinary approaches are needed for the solution of complex hydrological problems influenced by processes traditionally assigned to other disciplines. University curriculae need to be re-evaluated to provide interdisciplinary background and a common language for problems formulation.

It is desirable to analyze the credibility of hydrological methods and results, starting, for example, from critical re-evaluation of assumptions upon which hydrological methods are based. One should be aware of weaknesses of several established practical tools in hydrology, where extrapolation is used.

Acknowledgments. The author is indebted to Professors J.C.I. Dooge, University College Dublin and Galway and R.J. Sterrett, Colorado School of Mines for their valuable comments.

References

Burges, S.J., Trends and directions in hydrology, Water Resources Research, 22, 1S-5S, 1986.

Dooge, J.C.I., On the study of water, Hydrol. Sci. J., 28, 23-48, 1983.

Dooge, J.C.I., The waters of Earth, Hydrol. Sci. J., 29, 149-176, 1984.

Dooge, J.C.I., Scale problems in hydrology, Kisiel Memorial Lecture, Department of Hydrology and Water Resources, University of Arizona, Tucson, February, 19th, 1986, in press, 1987.

Eagleson, P.S., The emergence of global-scale hydrology, Water Resour. Res., 22, 6S-14S, 1986.

Klemes, V., The unreliability of reliability estimates of storage reservoir performance based on short streamflow records, in: Reliability of Water Resources Management, Water Resour. Publ., Fort Collins, Colo., 193-205, 1979.

Klemes, V., Dilletantism in hydrology -- Transition or destiny?, Water Resour. Res., 22, 177S-188S, 1986.

Klemes, V., Hydrological and engineering relevance of flood frequency analysis, presented at the Intern. Sym. on Flood Frequency and Risk Analyses, Baton Rouge, La., May 14-17, 1986a.

Kundzewicz, Z.W., L. Gottschalk and B.W. Webb (editors), Hydrology 2000 Report, IAHS Publications, 1987 (in press).

Linsley, R.K., Meeting of the AGU Committee on Status and Needs in Hydrology, Trans. AGU, 693-698, 1964.

Meier, M.F., Snow and ice in a changing hydrological world, Hydrol. Sci. J., 28, 3-22, 1983.

Roederer, J.G., Tearing down disciplinary barriers, EOS, Trans. AGU, 66, 681+684-685.

References to Quo Vadimus Symposium extended abstracts

International Union of Geodesy and Geophysics, A collection of Contributions to IUGG Symposium U1 -- Quo Vadimus -- Where are we going?, dedicated to the 100th anniversary of the birth of F.A. Vening-Meinesz, XIX General Assembly, Vancouver, Canada, Aug. 9-22, 1987, preprinted manuscript:

Duckstein, L., H.P. Nachtnebel and I. Bogardi, Long range risk analysis: Uncertainty and fuzziness, p. 157.

Krausneker, P., Hydrology as a contribution to water management marked by a high sense of responsibility, pp.153-156.

Lliboutry, L., Rheology of metamorphic rock ice, pp. 158-159.

Rango, A., Hydrology, pp. 148-149.

Ubertini, L. and L. Natale, Integrated formulation of hydrological models and design of hydrological monitoring systems, pp. 150-152.

STATEMENT TO FOLLOW QUO VADIMUS - HYDROLOGY BY Z.W. KUNDZEWICZ

Mark Meier

Institute of Arctic and Alpine Research
Department of Geological Sciences
University of Colorado, Boulder, Colorado 80309

Professor Kundzewicz has given us a well-reasoned overview of hydrology today, and he mentions some of the major problems that are likely to confront us in the future. He dwells mainly with philosophical concerns within the field today, and I fully agree with his assessment of such concerns as: the schism in hydrology between the desire to advance the basic science versus the engaging pressures to solve immediate problems, the problems of scale and aggregation, the emerging recognition of the need to take a more global view of hydrology including the oceans, the difficulties inherent in working in such an interdisciplinary environment, and the credibility problem involved with the use of existing empirical methods and models. Many of these concerns relate to hydrology as a study of water resources, but the field is broader; it is a basic, global, environmental science.

It is difficult in a short paper to summarize the present and future state of a broad and vigorous area such as the hydrological sciences. This is particularly true because of the interdisciplinary nature of hydrology. It would be far simpler to discuss where we are going in groundwater modeling, or in flood analysis and prediction, or in soil water biochemistry, or in glacier dynamics, or in any one of many other relatively compact disciplines. Thus Professor Kundzewicz' paper tends to be rather general and philosophic in tone, and my remarks must be of a similar vein.

First, we should emphasize the central importance of the hydrologic sciences in emerging concerns about the future global habitability, the interacting of the geosphere and the biosphere. Hydrologic processes are critical in biogeochemical cycling, are essential to and often regulate the terrestrial biosphere, play a significant role in the global climate machine, and contribute to the record of environmental change that gives us perspective in our attempts to understand global change. Critical questions for the future include:

1) How can we measure the global hydrological cycle, including precipitation, evaporation, transpiration, snow and ice melt, and surface and groundwater flow, both regionally and globally, including land and ocean?

2) Can we define and understand the chemistry, biology, and physics of the precipitation/evaporation, soil moisture flow, atmospheric chemistry, and biogeochemical fluxes in different environments?

3) How will the above processes be affected by mankind's actions, modifying landuse and changing air, soil, and water chemistry?

We need answers to these questions, in order to develop models of global biogeochemical cycles and climate that can be used with confidence to avoid or minimize detrimental impacts to society.

Second, we should note that hydrology suffers, in almost all aspects, from a severe lack of observational data. This is due in part because many hydrologic processes vary rapidly with time and space requiring an extensive, and thus expensive, data collection effort, and in part because the science itself and some of its objects of study, such as anthropogenic pollutants, are continuously evolving so that new data collection programs must be developed. A significant, aggravating factor is the difficulty of funding monitoring programs in the service of science. These programs cannot promise the scientific "breakthroughs" that are so important to most funding agencies, and yet useful monitoring requires a stable, long-term source of funding. Most long-term data sets in hydrology were collected for practical reasons, but the practical needs do not always produce the appropriate data for advancing basic understanding. Some questions include:

1) Can satellite sensors be developed that can

measure near surface hydrologic properties such as soil moisture, in spite of the complications of vegetation, darkness, relief, and cloud cover?

2) How can we map the time evolution of precipitation patterns, especially the intense but areally-restricted convective storms?

3) Can we sense, from satellite altitudes, vapor fluxes and biogeochemical fluxes in the atmosphere?

4) Is there some way to assure the continuity of vital, long-term monitoring programs, in both developing and developed countries?

Lastly, we should note that the science of hydrology is largely based on studies done in the semi-humid north temperate latitudes, but hydrologic processes are different, both qualitatively and quantitatively, in other parts of the world. How useful is a rainfall-runoff model in the Arctic, or a flood-frequency analysis in the desert? Wetlands, high mountains, rainforests and cloudforests, and ice sheets are other environments which have their own special hydrologic problems, problems that are often given only token mention, or none at all, in hydrologic texts. Yet we now need to understand hydrologic processes in these unusual and sometimes extreme environments, both because of local needs and in the service of global hydrology. The critical question for the future is simply, how can we support and sustain basic hydrologic research in extreme and unusual environments, bearing in mind that the driving force of economics is weaker in these environments, and that these hydrologic processes and effects, important as they are to Earth-system science, cannot be extrapolated for use in the more populated areas of the world.

STATEMENT ON QUO VADIMUS – HYDROLOGY

Peter S. Eagleson

Department of Civil Engineering
Massachusetts Institute of Technology, Cambridge

Introduction

Kundzewicz [1990] has discussed the growing importance of large scale problems in hydrology and the need for interdisciplinary approaches to them both in research and in education. He does not give many examples of critical research frontiers, however. This contribution will attempt to illustrate with examples what many believe to be the critical issue facing hydrology today—the need to step outside the traditional applications oriented view of hydrologic science to one which is geophysically based.

The Problem

Since the beginning of civilization the development of hydrology has been largely in the hands of engineers working on the classic problems of water supply, food production and the reduction of natural hazards. Their considerable success at these tasks is evidenced by the high standards of public health and safety enjoyed by the urban populations of the more developed nations. Nevertheless, the pragmatic focus of the field has, until very recently, retarded the development of fundamental hydrologic science in comparison with the earth, atmospheric, and ocean sciences. This has resulted in a scientific and educational base that is inadequate for solution of many emerging problems arising from the accumulated effects of man's active presence on this planet. While there are certainly many current problems in applied hydrology for which the existing science base is adequate, the prognosis of global change [National Research Council, 1986] suggests the need to give priority to building hydrologic science.

Long-term maintenance of Earth's health requires understanding its geophysiology and metabolism. This relatively recent acceptance of our planet as an evolving living system is altering the perspectives of most geosciences, and hydrology is no exception. Water is both the blood and the lymph of this body and there is renewed interest in their common circulatory system, the hydrologic cycle. Hydrologists are having to consider the atmosphere, land surface, and subsurface aquifers as a coupled water system interactive on time and space scales inclusive of global-scale processes. In so doing, we are having to draw increasingly on the fields of meteorology, geology, physics, chemistry, and biology. This is revealing some important deficiencies in our knowledge of water science:

Some Research Opportunities in Hydrologic Science

- *Global Moisture Cycle* – Our quantitative knowledge of many of the reservoirs and fluxes is surprisingly poor, especially for those extreme climates of the world without a history of systematic surface observations such as the oceans, arctic and alpine regions, deserts, and much of the tropics. This is particularly true for oceanic precipitation and evaporation, long-term changes in land ice reservoirs, and short-term changes in seasonal snow cover where new measurement technologies are needed. Modern geochemical instruments enable measurement of isotopes and rare gases at concentrations which promise safe and inexpensive tracer definition of water sources and pathways.

- *Mesoscale Land Surface Behavior* – The spatial integration of heat and moisture fluxes over heterogeneous mesoscale river basins and grid squares of numerical models is fundamental to the role of these interactive atmosphere/land surface elements in the regional and global climate. How do we represent the average hydrologic and vegetal behavior of such systems? Coordinated observations at this scale are just beginning to be made [Andre et al., 1986; Rasool and Bole, 1984; DOE, 1986].

- *Fluvial Geomorphology* — What is the dynamic relationship of a stream channel network to the landform, to the fluxes of moisture, heat and sediment, and to the growth of vegetation? Is there an equilibrium here that holds the key to dynamic similarity? If so, how sensitive are these equilibrium states to anthropogenic changes? The principles of hydrologic similarity or scaling are unknown and yet important to the kind of comparative hydrology necessitated by a sparse global data base.

- *Precipitation Fields* — Rainstorms are a critical factor in the dynamics of the hydrologic cycle and therefore prediction of their space-time characteristics is essential to hydrologic forecasting. Hope for improving the accuracy and lead time of forecasts depends upon incorporating in hydrologic models more of the complex combination of the fluid dynamic, thermodynamic, and cloud microphysical processes that provide the space-time correlation in rainfall fields. The coordinated mesoscale observations on which such advances will be based have not been made.

- *Soil as a Reactor* — The topmost meter of soil is a chemical and biological reactor, and the water and solutes stored there play central roles in coupling the various biogeochemical cycles. The interaction of both water and dissolved chemical species with the soil medium and its biota is important to element cycling and to the fate of toxic spills.

- *Stream Chemistry* — Geochemical characterization of the surface and subsurface contributions to river flow as a function of space and time should provide improved understanding of element cycling and soil leaching.

- *Geohydrology* — Quantitative understanding of the role of deep ground water in tectonic processes, in subsidence, and in the transfer of heat and mass in crustal rock will contribute to our forecasting of earthquakes and our exploitation of geothermal energy.

We notice two characteristics of these critical research areas. All demand field experimentation, for hydrology is currently constrained by the lack of coordinated data sets, and all are heavily interdisciplinary, involving meteorology, geology, physics, chemistry and biology, for here are the frontiers of the science.

Educational Programs

The growth of hydrologic scientists cannot occur efficiently in educational programs dominated either by applications-oriented constraints or by predominant undergraduate preparation in engineering. A thorough background in mathematics, physics, chemistry and the geosciences is needed.

Acknowledgments. Much of the above material appeared earlier in a newsletter of limited circulation [Eagleson, 1987] and other has come from internal memoranda of the Committee on Opportunities in the Hydrologic Sciences, a committee of the Water Science and Technology Board, National Research Council. This committee is charged with defining the hydrologic sciences, assessing the current state of their development and of their couplings with related geosciences and biosciences, and identifying promising new frontiers and applications opportunities along with the appropriate framework for education and research.

References

Andre, J. C., J. P. Goutorbe and A. Perrier, HAPEX-MOBILHY — A hydrologic atmospheric pilot experiment for the study of water budget and evaporation flux at the climatic scale, Bull. Am. Meteor. Soc., 67, 138-144, 1986.

Department of Energy, Program plan and summary — Remote Fluvial Experimental (REFLEX) Series, DOE/ER-0254, 1986.

Eagleson, P. S., Problems and opportunities in the hydrologic sciences, Water Sci. and Tech. Bd. Newsletter, 4(3), National Research Council, Washington, D.C., May, 1987.

Kundzewicz, Z. W., *Quo Vadimus* — Hydrology, (this volume).

National Research Council, U. S. Committee for an International Geosphere-Biosphere Program, Global Change in the Geosphere-Biosphere, National Academy Press, Washington, D.C., 1986.

Rasool, S. I. and H.-J. Bolle, ISLSCP: International Satellite Land-Surface Climatology Project, Bull. Am. Meteor. Soc., 65(2), 143-144, 1984.

THE LITHOSPHERE OF THE EARTH AS A LARGE NON-LINEAR SYSTEM

V. Keilis-Borok
Institute of Physics of the Earth
Moscow

Background

The questions formulated here concern the dynamics of the lithosphere. We look for its integral comprehensive representation, which would encompass at least the following major phenomena.

The lithosphere has discrete hierarchial structure. It consists of a hierarchy of volumes ("blocks") divided by less rigid boundary zones. The largest volumes are the tectonic plates which are divided by fault zones; the smallest volumes are the grains of rocks or crystals of minerals divided by interfaces. A boundary zone, at least of a high rank, presents similar hierarchy: it consists of volumes, divided by boundary zones etc.

The volumes move relative to each other against the forces of friction and cohesion in boundary zones, which bind them together. In seismically-active regions a significant part of the movement is realized through earthquakes. Friction and cohesion in turn are controlled by many mutually dependent processes: by interaction with fluids; by different mechanisms, e.g. Rhebinder effect (stress corrosion) and lubrication; phase and petrochemical transitions; fracturing and pseudoviscous flow etc. Each of these processes generates strong instability and can quickly reduce effective strength by the factor up to 10^6 in some cases.

The mechanical strength of the volumes themselves may be affected by most of these processes and some others such as buckling. All of them are influenced in turn by heat flow within and through the system.

It is possible that none of these processes can be singled out, as a major one, which would permit the others to be neglected in some approximation. Therefore the formulation of equations, directly depicting each of them and their interaction is not realistic and probably not even possible in principle. Hence the problem arises -- to design a theoretical model, which depicts directly the gross, integrated traits of lithospheric dynamics. If it is possible, it has to be at the cost that the details below some level are not retrievable by the model (as Brownian motion in gas dynamics). A partial solution to this problem is given by the models of mechanics of continuous media, with the values of parameters replaced by "efficient" ones (e.g. decrease of elastic modulus representing micro-fracturing). This traditional approach gives valuable results but it is applicable only to a limited range of problems.

The study of a more general model of the lithosphere's dynamics is so far at the very preliminary stage, namely the search for empirical regularities. The questions below are formulated accordingly. Also allowed for are the evidences, so far inconclusive, for deterministic chaos and self-organization in the lithosphere's dynamics on time-scales no less than 10^2 years.

Problems

1. On the integral characteristics of an earthquake sequence (defined as a function of time at a sliding time window within a certain region and energy range).
   -- For what functions is there a self-similarity or partial self-similarity regarding the energy and/or space?
      Is it limited to a certain range of: energy; size of a region; tectonic environment?
      What ratio of time, space and energy scales (e.g. what normalization in time, space and energy) is required for such self-similarity?
   -- What is the minimal set of functions, which determine the basic traits of seismicity, including the approach of a main shock (in time scales, say, 0.1, 0.01, 0.001 of its average recurrence time)?
   -- What are the relations between these functions ("invariants")?
      What is the type of transition to chaos (if any)?

2. Similar questions on the integral characteristics of tectonic movements, on the time-scales say 10^4, 10^6, 10^8, years.
   -- Does self-similarity (or partial self-similarity) exist, and if so, what are its limits?
   -- What is the minimal set of functions, required to depict the basic traits of tectonic history; in particular -- can this history be described as an intermittent sequence of instability episodes?
   -- What are the relations between these functions?
   -- Does chaotic behavior occur?

3. What is the fractal dimensionality of systems of faults; of clouds of epicenters and hypocenters?

International Union of Geodesy and Geophysics

Contributions to the Quo Vadimus Symposium

Dedicated to the 100th Anniversary

of the Birth of

F.A. Vening Meinesz

XIX General Assembly, Vancouver, Canada

August 9–22, 1987

CONTRIBUTIONS TO THE QUO VADIMUS SYMPOSIUM

George D. Garland
Vladimir I. Keilis-Borok
Helmut Moritz

Over seventy short contributions were submitted to IUGG Symposium U1 "Quo Vadimus: Where are we going?" Some of the original contributions have been expanded and published elsewhere in the volume. One or two later contributions have been added to the "green book" distributed to symposium participants. The following pages contain this collection of short contributions that answer our question "Quo Vadimus?"

The aim of the symposium was to idenify major key problems of geodesy and geophysics for the next few decades.

An encouraging precedent was the famous endeavor of David Hilbert, who in 1900 formulated a set of 23 most important mathematical problems "which contemporary science poses and whose solution we expect from the future." Hilbert's problems have been highly influential in the subsequent development of mathematics. Today, one man's effort may no longer be sufficient to achieve such a goal, especially in highly complex and controversial areas of Earth sciences. Still, we hope that the success of Hilbert's list of problems may serve as an encouragement to our collective endeavor.

The contributions, each of which is expected to contain one or several clearly formulated open problems, are arranged in such a way that, as far as possible, contributions on related subjects can be found somewhere near each other, beginning with general problems of the Earth as a planet in the solar system and then gradually going from geodesy and gravity to the physics of the solid Earth and to geomagnetism, finally passing from the ionosphere to the atmosphere and the hydrosphere.

The summaries are reproduced in the original form submitted by the authors, with only the most superficial editing, in order to respect the authors' individuality. Duplications and overlaps will be natural and should even be interesting and fruitful.

Geophysical Monograph 60
IUGG Volume 10
©American Geophysical Union

Academician E.K. Kharadze

Abastumani Astrophysical Observatory
Academy of Sciences of Georgian SSR

SOLAR-TERRESTRIAL CONNECTIONS

1. What is the mechanism of the solar activity action on the processes in the Earth's atmosphere?
2. What is the mechanism of interrelation of the Earth's atmosphere separate layers?
3. What is the reason of regular oscillations of the terrestrial atmosphere structural parameters and the mechanism of their connection with regular processes in the heliomagnetosphere?

EARTHQUAKE PRECURSORS

1. What is the nature of the effect of the processes in the terrestrial crust on the variations of the upper atmosphere parameters at the period of the earthquake preparation?

ATMOSPHERIC AEROSOL

1. What is the nature of the mesosphere aerosol layer with the concentration maximum at the height of about 50 km? How is it spatially distributed and what is its time behavior like?

The available limited data seem to show no correlation of its behavior with the upper and lower layers, the formation of which is connected with the matter influx of meteor and volcanic origin correspondingly. This gives ground to believe the mesospheric aerosol formation as a result of intra-atmospheric processes. If so, then a comprehensive study of the mesosphere layer would help in a better understanding of the transformation mechanisms of the atmospheric constituents, in particular, sink mechanisms of admixture at the height of 40-60 km.

EARTH MASS -- STABILITY AND FLUCTUATIONS

Cdr. C. Copaciu
Hydrographer, Ret.
7 rue Bel Respiro, Monte-Carlo, MC-98000 Monaco

The complete description of a geodynamic processus within the Earth is at present beyond any other grasp. However, a comprehensive theory could be developed to contribute to the understanding of some points of interest (e.g., mantle mass transport) to follow their evolution and to predict their future behavior. What is evidently needed for the complete formulation of the problem, is the pertinent value of some information about initial conditions, the distribution and evolution of physical parameters, and the constraints. The energy balance must be described, as well as that of the different "engines" that drive motions, i.e., the transport of mass and energy, the geometry of the flow, the gravity field, the rheological state of mantle materials, etc. To solve these problems, a systematic approach, coupled with an elegant and extensive mathematical attack, must be applied.

As understanding increases, it may be possible to obtain an accurate, global dynamical overview of the Earth's dynamic phenomena. However, many questions, and not of the least importance, remain open. Several of them are summarized hereafter.

A. Geodynamics Uncertainties

1. What is the relation between the orientation of the mantle relative to the spin axis and dynamical processes?
2. Is the reorientation of the Earth responsible for the stress that initiates important modifications in the mass movements?
3. What is the relation between the mantle processes and the rotational dynamics of the Earth?
4. Is the cooling of the upper mantle responsible for the geoid's low active subduction and continents' immobility which reorient the mantle relative to the spin axis?
5. What is the interaction of tides and mass transport?
6. What can be the impact of large bodies on the Earth's interior?
7. What is the relation between polar wandering and mass movements?
8. Is temperature of a secondary importance in the mantle activity? Is it determined by or determining the mantle activity?
9. Do temperature and pressure-dependent viscosity play a role in the lineated patterns of gravity in the upper mantle? Are they important in several situations involving deep circulation?
10. Is the differential/separation the dominant large-scale process in the mantle?
11. Is melting one of the most important driving forces in mantle convection?
12. What are the distribution and topology of heterogeneities in the mantle? What is the time survival of mantle heterogeneities?
13. Is the mantle compositionally stratified, and is its convection confined to its layers?
14. What is the source of isotopic anomalies within the mantle? Is it the material of adjacent lower layer, the large scale delamination of the continents, or the breaking up of large convective cells?
15. Is the transition zone ($\simeq$670 km) a region of phase changes, of chemical discontinuity, or a compositional zone?
16. How can the seismic properties of the mantle, its bulk chemistry, and the isolation of geochemical reservoirs be explained?
17. Are the geochemistry, morphology, and timing of igneous processes affected by solitary wave instabilities arising in the viscous mantle?
18. Does the oxygen fugacity, together with temperature and pressure, control the petrogenesis of mantle-derived magmas?
19. How are the effects of the composition and speciation of the mantle fluids an initial input in the evolution of the Earth's crust, mantle mass, and hydrosphere?
20. What is the relation of Sr. and Nd. isotopes with the evolution of the mantle reservoirs?
21. What are the links between short-scale convection instabilities of short wave-length and pressure-dependent viscosities in the mantle convection?
22. Can the unstable viscous fingers break up in the mantle branched structures?
23. Are there some implications of fractal geometry on random structures in the mantle, or on growth processes in the Earth, or on mantle properties near critical points?
24. How can the fingers instability in the interface between two mantle fluids with different viscosity be understood?
25. Are dynamic processes in the mantle transient or steady states?
26. Are the fluctuations produced by two competing energy sources -- thermal convection and growth of the inner core -- the cause of magnetic field?
27. Is there within the Earth a poroclastic shell which could assume two states: "inflation" or "deflation"? Is it affecting the polarity of magnetic field?
28. How can the coincidence in time between a high rate of magnetic reversal spurts, tectonic activity, and body impact be explained?
29. What is the relation between resonance of Earth's magnetic field and mantle phenomena?
30. Why is the transition field not axisymmetric, and why is it characterized by rapid fluctuations?

B. Mathematic Tools

1. How can the fluctuation theory be applied to non-linear, far-from-equilibrium systems?
2. Is it always possible to solve the equations with multiple degrees of freedom in the fluctuation theory?
3. What is the mathematical formalism which gives a true idea of the behavior of a non-linear system? Does such a mathematical formalism exist?
4. Is it possible, in the complex dynamic and spatial theory to take no account of non-linear evolution, but utilize a sophisticated analysis for the systems with little degree of freedom?
5. Can the introduction of the "sensitivity" (feeble dynamic stability) of an ensemble of dynamical trajectories solve the uncertainty of Earth's prior state or states?
6. Are the domain of validity and content application of the dynamic Earth system certain?
7. Is it possible, in the Earth fluctuation theory, to change from systems with little degree of freedom to systems with great degree of freedom, and reciprocally?
8. Can the Earth fluctuations be explained by the mixing of complex dynamic effects with the effects of complex spatial structures?
9. What is the role of the theory of stability and the non-equilibrium Earth dynamic system?

10. What happens on the dynamic branch?
11. What happens beyond the instability of the dynamic branch?
12. What is the velocity of growth of new structures?
13. What is the type of solutions of differential equations on bifurcation or critical points?
14. Can the distribution probability and the short-time evolution of fluctuations be analyzed?
15. Is it more convenient in the convection to use, instead of the Navier-Stokes equation, the Lorentz model, or the non-linear stochastic equations, or the Fokker-Plank equation, or the "master" equations, or others?

C. Comments

It seems that mathematics to deal with geophysical problems have not been invented . . . mathematicians preferring to deal with generalities rather than setting up a mathematical system of physical laws to correlate given empirical phenomena. However, new ideas and concepts in physics, physical chemistry, biology, etc., have, since recently, given rise to new developments and orientation in mathematical approach to provide more rigorous tools for further developments. The burst of new theories and methods -- the catastrophe theory, fractal geometry measure theory, strings, twisters, physical geometrization, etc. -- to relaunch new reflections about certain paradoxes, could suggest an alternative to different assumptions about the Earth mass dynamics.

What are the mathematical and physical tools which may be used to carry out studies of the geodynamic processes of the Earth systems?

The basic postulate is that all the phenomena of the Earth systems may be studied with reference to both the laws of physics appropriate to the specific non-linear interactions to be considered, and the far-from-equilibrium conditions. Furthermore, the very fact that is interesting in modeling the macroscopic phenomena, is to introduce a more drastic simplification, permitting to adopt a contracted description in terms of a limited number of non-linear parameters. A physical system is inevitably subject to perturbations of various kinds. For the Earth systems, these are external excitations of its environmental conditions, and internal fluctuations generated by the Earth system itself.

As a result, the Earth system continuously deviates -- though usually weakly -- from the balance equations of macrovariables and dynamic microvariables, that I. Prigogine *et al.* have established in an elegant generality linking non-equilibrium thermodynamics with statistical mechanics, at least in the linear region. However, the theory of the "dynamics of feeble stability states" expresses that the dynamic systems develop macroscopic phenomenology, whereas the little quantum actions limit the sensible consequences to the microscopic level. Again, the classical dichotomy reappears, giving place to simplicity and complexity, order and disorder, determinism and hazards, conservation and dissipation, linear and non-linear, etc.

Thus, from a local concept described by mathematical analysis, the Earth dynamics could evolve in a global concept, described by geometry and topology.

ATMOSPHERE-EARTH-OCEAN DYNAMICS

J.T. Kuo
Columbia University, New York

A. Introduction

The precise knowledge of the ocean and earth dynamics everywhere on the globe, including the continental shelves, slopes and deep oceans, is of fundamental importance in physical and satellite geodesy, astronomy, geophysics, meteorology, as well as space and ocean technology. As the interplay of earth and ocean dynamics increasingly takes place, we will more and more turn towards time-dependent aspects of the studies. It is equally apparent that the seemingly minute influence at one point, of various phenomena of the atmosphere on both the earth and ocean tides, must be brought into play. We are not only dealing with the earth-ocean system, but the "Atmosphere-Earth-Ocean" system, as a whole.

B. Key Questions

1. How do we realistically model a time-dependent response of an elliptical, rotating, laterally inhomogeneous, inelastic Earth, with the realistic Earth interior, oceans and continents, and enveloped by the atmosphere due to not only the tidal generating forces of the Moon and the Sun and the non-periodic forces of meteorological nature, but also the various body forces within the Earth, notably the convection forces?
2. What are the dynamic effects of the core on the tidal and load deformation of the Earth, and particularly, how are the mantle and the outer core, as well as the inner and outer cores of a realistic Earth, as referred to above, coupled in the light of reexamining the Adams-Williamson adiabatic and the Adams-Williamson-Birch non-adiabatic approximation?
3. Is the Atmosphere-Earth-Ocean Dynamic System in any way related to theories of the geodynamo?

GLOBAL WATER AND HEAT EXCHANGE

S.G. Dobrovolski
Institute of Water Problems
USSR Academy of Sciences, Moscow

1. Whether the natural redistribution of water mass between the World ocean and the land, at the time scale of years, dozens of years and, perhaps, centuries, occurs as the Wiener process? Is the Hasselmann's law of climatic variability without feedback ("w^{-2}" law) applicable for the description of the water mass redistribution between ocean and Antarctic ice sheet? Between ocean and each continent?
2. What is the main mechanism of the longtime (month-to-month and year-to-year) changes of ocean surface temperature anomalies and evaporation anomalies? (Sea surface temperature is the most important oceanic

parameter which influences the rate of evaporation from the surface and thus is situated in the very beginning of the global water exchange chain).
3. If the Markov presentation of the atmospheric water content changes and white-noise presentation of the atmospheric water transport variations is commonly applicable for the "medium" -- between synoptic and climatic -- time scale? For the climatic time scale?
4. What is the geographical distribution of the integral correlation scale (correlation time) of interannual runoff variability? What is the mechanism of the year-to-year runoff inertia -- the main feature of the long-term river runoff variations?
5. What effect with emphasis on the water mass redistribution in the system "ocean-atmosphere-continents" may be produced by the greenhouse gases increase in the atmosphere? Which mechanism will prevail in the case of global man-made heating: ice sheets melting or their extension due to a possible increasing of precipitations over ice sheets?

DEVELOPMENT OF GEODESY AND GEOPHYSICS IN AFRICA (EAST AFRICA CASE)

A.L. Mutajwaa
State Mining Corporation, Box 981
Dodoma, Tanzania

A. Introduction

The introduction of the geodesy lecture program at the University of Nairobi under Prof. Lars Asplund in early 1970 was indeed a good sign of the beginning of harnessing geodesy potential in this part of Africa.

Since then, there have now been around fifteen young geodesists almost evenly distributed between Kenya, Tanzania and Uganda, all at M.Sc. level and above. Tanzania has two at M.Sc. level and three at Ph.D. level.

The IUGG in collaboration with other agencies established the Commission of Geodesy and Geophysics for Africa in mid 1970 and this led to some few symposia held on geodesy and geophysics in Africa. Other attempts have been to create regional centers such as the Nairobi Regional Centre for Mapping and Remote Sensing, to try to fill the gaps.

Unfortunately, we have not lived to the expectations of Prof. Asplund and his colleagues. Our young geodesists (and few geophysicists that there are) have had no opportunity to translate the immense knowledge acquired in the West to develop and use it in the development of their nations and the disciplines themselves.

B. Key Questions

1. What course of action could be taken by IUGG and other related agencies to develop geodesy and geophysics in Africa?
   a) Could IUGG assist in securing the needed geodetic and geophysical equipment needed for acquisition data and computation thereof? What about their maintenance. What arrangements could be made to meet the expenses?
   b) In view of the diversity of politics, how could the IUGG and related agencies assist in setting up common modern (computer) computing facilities for geodesy and geophysics should these warrant such facilities? Alternatively, medium computing facilities may be availed at national levels for nationals engaged in these disciplines: e.g. Ardhi Institute in Dar-es-Salaam in Tanzania.
   c) How could IUGG and related agencies assist in retrieving some data available in other countries outside the countries of origin of those data, assuming that other factors have been taken care of, e.g. settlement of certain obligations between country of origin and the outside country with the data? This could help in creating some data banks for geodesy and geophysics.
   d) In view of the foreign exchange constraint, the geodetic and geophysical communities in Africa are rarely up to date with the development of these disciplines in the developed communities. How could IUGG and related agencies assist in this information dissemination to Africa?
2. With the knowledge of the tremendous available resources in many parts of Africa and also the available agricultural potential, how could IUGG and related agencies assist in extending geodetic and geophysical control networks in Africa, in addition to the well known 30th Meridian Arc, the 12th Parallel, and to lesser extent, some few added national networks either conventionally or satellite fixed, to help exploit these resources?
   a) How could the IUGG and related agencies assist in the realization of the idea of having a common African Geodetic Datum; so that efforts begin to readjust African geodetic (and geophysical) networks?

GEODESY, GEOPHYSICS AND REMOTE SENSING

Chief Administration of Geodesy
and Cartography under the Council of
Ministers of the USSR

Key Problems

1. Study of temporal variations of the Earth's figure, size, outer gravity field and sea level surface.
2. Preparation of reliable Earth's interior structure model and determination of relations between geophysical fields variations, surface deformations and processes inside the Earth.
3. Creation of measuring systems that operate on different principles of physics intended for real-time high-accuracy positioning and geophysical fields determination.
4. Mapping, geodetic and geophysical study of bodies in the solar system.
5. Preparation of static and dynamic (dynamic means variable, depending on day, season and meteorological conditions) models of optical characteristics (brightness, threshold and space-frequency characteristics) of terrain at local and global levels.
6. Elaboration of theory and methods to determine

informational content of each of specified n-characteristics and any of their combinations when solving problems of classifying a specified object classes alphabet.
7. Elaboration of the most general constructive theory of planes and three-dimensional spaces representation to solve applied problems and reveal geometric essence of space in the Universe.

EXPERIMENTAL GRAVITY RESEARCH

H.-J. Treder
Einstein Laboratory of Theoretical Physics
Academy of Sciences of the GDR

All deviations from the theory of gravitation by Newton and Einstein touch fundamental problems of present physics and should be examined in experiments. (Some contributions of geodesy, geophysics, space research and astronomy aimed at a solution of these problems are being discussed by M. Steenbeck and H.-J. Treder in the booklet "Possibilities for experimental gravity research," Akademie-Verlag, Berlin 1984.) In any case, the key problem is to systematically analyze and exclude all effects which are calculable in the frame of the classical theory of gravitation.

There are the following problems:

1. The absorption of the gravitational flow by matter, especially by the terrestrial body (according to Laplace, Seeliger or Majorana).
2. The self-absorption (suppression) of the gravitational field by the gravitational potential; those are changes of the active gravitational masses depending on the local value of the gravitational potential (Einstein-Cartan-theory).
3. Variation of the gravitational constants in space and time (Diracs hypothesis).
4. Variations of the relation of inertial and gravitational mass (in the sense of the Mach-Einstein-doctrine).

These questions lead to the verification of the existence of non-Newtonian tides and of secular variations of gravity as well as to the possibility that gravimeters working on different physical principles (free fall or spring balance, respectively) yield different results of measurements. Most of these hypothetic effects reach theoretical orders of some 10^{-8} up to 10^{-7} gal of different periodicity.

GEODESY AND GRAVITY

S. Stefanescu
Romanian National Committee of Geodesy and Geophysics
R-70201 Bucharest-37, Romania

1. Impact of Relativistic Mechanics on Geodesy

Given the remarkable recent progress both in observational geodesy and in its theoretical infrastructure, on the one hand, and the new outlook of mechanics on the conceptual framework of fundamental natural phenomena, on the other, much is to be expected from a thorough integration thereof, perhaps rather from their intimate synergism.

2. Structure and Evolution of the Gravitational Field

Spatial distribution and temporal changes (secular variations?) of this field need both refined detailed observations and faithful theoretical structuring wherefrom along with more accurate knowledge a better understanding is also to result.

GEODESY

E.W. Grafarend
Department of Geodetic Science, Stuttgart University
Keplerstrasse 11, D-7000 Stuttgart

1. Relativistic Reference Frame

For high precision geodesy the re-establishment of a precise geocentric reference frame is needed which is accurate in nanoseconds with respect to time and millimeter with respect to space. Such a geocentric reference frame has to be based on special and general relativity, namely the Einstein theory of gravitation.

Since modern geodetic instrumentation, namely satellite and terrestrial clocks, depend very much on a nanosecond time reference frame this problem is considered of key importance. In as far as the geopotential reference surface or equipotential surface is free of relativistic effects, it has to be investigated as well as the Bjerhammar proposal of relativistic levelling. For positioning an open question is how to relate a geocentric tetrad with the observer's tetrad on the Earth's surface. A relativistic definition of the mass center of the Earth has to be found.

2. Secular Variations of the Terrestrial Gravity Field

The representation of the gravitational potential of the Earth in terms of scalar spherical harmonics as base functions is standard. Since the Earth is not a rigid body, the spherical harmonic coefficients change in time. Measurements of the temporal change of the term of second degree have been performed. A key question is the influence of a time varying terrestrial gravity field on the Earth dynamics like precession, nutation, polar motion and length-of-day.

3. Mass Irregularities -- The Geoid

Within the determination of the geoid the mass densities between the geoid and the Earth's topography have to be precisely known. The key question is how accurate the geophysical prior information of mass densities has to be known in order to guarantee a one centimeter geoid, namely in areas of rough topography.

For the interpretation of the long wavelengths of the geoid the impact of convecting mantle and isostatic mass compensation has to be investigated, namely to get a better geodetic understanding of plate tectonics.

4. Nonlinear Adjustment by Differential Geometry

The determination of a terrestrial point in geometry and gravity space is based on measurements within geodetic networks, e.g. by GPS satellite or ground based techniques. The corresponding observational models are nonlinear and very often sensitive to the point of linear approximation. A key question which has to be investigated by differential geometric means is the local geometry of nonlinear observational equations. What is the tolerable curvature of the model space under which linearization can be performed?

5. Non-Uniqueness of Spherical Harmonic Expansion of the Gravity Field -- Backus Effect

Backus effect: a standard geodetic procedure is the determination of the Earth's gravity potential from measurements of the absolute value of its gradient. G. Backus (1970) has shown that this procedure is non-unique. This fundamental problem has to be investigated for the Earth's gravity field, especially the size of area of non-uniqueness.

References:

(i) G.E. Backus: J. Geophys. Res. 75 (1970) 6339-6341
(ii) L. Hurwitz and D.G. Knapp: J. Geophys. Res. 79 (1974) 3009-3013
(iii) D.P. Stern et al: Geophys. Res. Letters 7 (1980) 941-944
(iv) D.P. Stern et al: J. Geophys. Res. 80 (1975) 1776-1782

6. Are the Equipotential Surfaces of the Terrrestrial Gravity Field Convex or Not?

For airborne or satellite gradiometry -- used for high resolution of the terrestrial gravity field -- the convexity of equipotential surfaces is a fundamental assumption. Investigations of the areas where the gravity field is flat, namely in the space inside and outside the earth, have to be performed. The key problem is known as the geodetic singularity problem where an attempt is made to identify the areas of a concave gravity field.

PHYSICAL GEODESY

P. Holota
Research Institute of Geodesy, Topography and Cartography
250 66 Zdiby 98, Prague-East, Czechoslovakia

1. What is the quantitative effect of the non-linearity of the geodetic free boundary value problems estimated in terms of a maximum or other equivalent functional norm? Important application: How "good" the reference model of the external gravitational field and figure of the Earth must be in order that the error of the linear solution of the respective boundary value problem would be as small as desired?

2. What methods of mathematical physics are the most suitable to be used under the requirements of high precision geodesy (say centimeter geodesy) for an effective solution of (linear) geodetic boundary value problems in solution domains with highly complex boundaries given by real topography? The aim is to develop the mathematical apparatus to an adequate accuracy practically, including the theory of approximation error estimates, in order to make the best of the quality of the terrestrial gravimetric information and of other input boundary data in the determination of the gravitational field and figure of the Earth.

3. What is the necessary quality of knowledge of: (1) the mass density distribution in the Earth's interior; (2) the topography of the physical surface of the Earth in order that the shape of the geoid could be obtained with the desired accuracy directly, i.e. from its fundamental definition as the shape of a particular level surface of the Earth's gravity field?

THE GEOID

V.S. Troitskiy
Institute of Physics of the Earth, USSR Academy
of Sciences, Moscow

To measure a high-precision global pattern of the Earth's shape, its rotation and dynamics of the surface are studied using the methods of radioastronomy.

The following global experiment may be suggested for this purpose:

At present, the radioastronomy techniques elaborated allow determining the Earth's pole position, distances along the chords compatible with the Earth's radius (with accuracy of several centimeters), and also the period of rotation (with accuracy to 0.1 msec). The duration of measurements does not exceed 24 hours. To make such measurements tens of radiotelescopes with diameter no less than 25 m and appropriate equipment should be placed over the entire globe. About 10 radiotelescopes placed in various countries can now also be utilized. The system should provide for the continuous observation and in fact will control the pole migration and the Earth rotation.

SPACE GEODESY

W. Baran
Institute for Geodesy and Photogrammetry
University of Agriculture and Technology
Olsztyn, Poland

1. What is the time-varying nature of the positions of surface stations in the geocentric coordinate system? The development in space geodesy methods based upon the use of laser techniques, satellite inferometry and other new measurement techniques will lead to obtaining

accuracies within the range of centimeters in determining stations positions. With such a precise positioning taking under consideration the influence of both global in character geodynamic phenomena (movements of tectonical plates, continental drift, etc.) and the features local in character will be necessary. The net's configuration and the optimal choice of observation will decide on the possibility and accuracy of evaluation of the influence of these features.

2. What kinds and sets of observations are necessary to determine the model of the Earth's gravity field with the accuracy of centimeters? Such a precise model will make possible the determination of the gravity field temporal variation. The precision of determining the variation will depend on the choice of measurement techniques and on the distribution of observations (in the four-dimensional space).

GRAVITY FIELD

W. Torge
Institut für Erdmessung
Universität Hannover
Federal Republic of Germany

1. How can the gravity field be effectively used for modelling the Earth's surface and interior state and dynamics? This implies studies about the information contained in different field quantities and spectral parts, and about the relation to other geodetic and geophysical data. The necessary improvement of gravity field modelling at global, regional and local scales urges to new space missions, and to a more intensive use of terrestrial data sets, as well as to the solution of theoretical problems as combining different data including the definition and realization of a "cm"-level reference surface.
2. How can gravity variations with time contribute to the modelling of neotectonic processes? This implies the development of geophysical models for vertical mass shifts within the framework of 3D-models, and optimum models for the combination of gravimetric and height observations. Establishment of efficient global and local control networks for monitoring vertical mass shifts asks for the improvement of gravimetric and spatial techniques to the 10^{-9} to 10^{-10} accuracy level, including the development of methods to monitor and model environmental disturbing effects.

GRAVITY FIELD VARIATION

P. Biró
Department of Geodesy
Technical University of Budapest
Hungary

What are the possible inside sources generating non-tidal variations of the Earth's gravity field? The Earth's gravity field is most probably undergoing to time variations being of non-tidal origin. These can be generated by the displacement (or redistribution) of the Earth's inner masses. Time variations in geopotential (i.e. the variations in the spherical harmonic coefficients of the geopotential) can be determined by repeated geodetic observations.

The problem to be solved: what possible inside mass displacements can be found generating the determined time variations in geopotential? In the first step it would be also useful to find reasonable models for the estimation of the magnitude and global distribution of possible variations in geopotential.

EARTH SIZE CHANGES

J.B. Zielinski
Space Research Centre
Polish Academy of Sciences, Warsaw

A. Introduction

Considering the dynamic behavior of the planet Earth - - we cannot take for sure anymore the traditional fundamental assumption of geodesy about the constant dimensions of the Earth. As a matter of fact, such established phenomena like the tectonic plate motion, mantle convection, pole wandering, climate changes all are the evidence that process of the Earth formation and evolution is not finished, the powerful energy sources are active inside of it and the outer space environment is changing as well.

Traditionally, the silent assumption of constant fundamental parameters is being accepted in most of works leading to the determination of the Earth ellipsoid, the geoid, the Earth gravity field or the Earth figure in general.

Now, in consequence of growing precision of the global geodetic measurements, the notion of changing dimensions of the Earth should be introduced. Then, the number of questions arises, which must be answered by theoretical or experimental research.

B. Key Questions

1. Are the fundamental definitions of geodesy like Earth ellipsoid, its axes, Stokes parameters, equatorial gravity value, geoid, etc., defined precisely enough, that their temporary changes in turn could be defined?
2. What is the role, if any, of the reference system for the study of the Earth size variations?
3. What is the expected rate of changes of the Earth's dimension parameters, as estimated from the point of view of physics, geophysics, cosmology, etc.?
4. Do we have in hand some measurement data containing (or possibly containing) some evidence about the Earth dimension changes?
5. What are potential possibilities of different geodetic techniques for investigation of the Earth dimension changes?
    a) satellite geodesy?
    b) VLBI?

    c) gravimetry?
    d) terrestrial networks?
    e) tidal measurements?
    f) others?
6. Which theories might be affected if non-constant Earth dimensions are assumed? Earth satellite motion theory? Lunar motion theory? Earth rotation model? Tidal model?
7. Is some worldwide geodetic research program for establishing the Earth dimensions changes necessary or wishful?

EARTH ORIENTATION & REFERENCE SYSTEMS -- THE FUTURE

Dennis D. McCarthy
U.S. Naval Observatory
Washington, D.C. 20392

Within the next few decades we are sure to see important changes in the concepts and realizations of the coordinate reference systems used in geodesy and geophysics. The accuracy with which astronomical and geodetic measurements can be made already are challenging the definitions and concepts which have been adequate until now. These changes may, in fact, be one of the more important considerations in geodesy and geophysics in the coming years since the precise definition of reference systems will serve as the basis for all future physical measurements. The accuracy with which these measures can be made now and in the future will not only enable us to validate current theories concerning the interaction between geophysical phenomena within the Earth as well as on its surface and in the atmosphere, but permit even further advances in our understanding of these processes.

Astronomical reference systems have been, and continue to be, based on optical measurements of the positions of stars and solar system bodies. It has now become clear that, while the definition may provide a reasonably accessible reference frame now, it is no longer adequate to represent the observational precision that can be obtained with the most modern techniques. Radio interferometry and laser ranging have now clearly demonstrated superior accuracy. Improvements in computing power and the success of interplanetary probes have also permitted new accuracy in the location of solar system bodies. The old analytic theories are no longer capable of locating these objects within the observational precision of the modern observational techniques. Modern, numerically integrated orbits have made the current definition of the ecliptic obsolete. With the validity of the definition of the ecliptic questioned, we must then reconsider the definition of the equinox. Perhaps a more suitable point for a zero of right ascension will come into use replacing the "First point of Aries." Similarly, the definition of the fiducial point in "inertial" space which we use to measure the Earth's rotation angle needs to be reconsidered, and perhaps the very definition of Universal Time time will be changed. These questions must be answered in the future, and the improvements in observational precision will not permit delays.

Geodetic reference systems have been based on conventionally adopted concepts and constants including the Conventional International Origin (CIO) and the BIH zero of longitude. The CIO serves as the origin for the description of the motion of the rotational axis with respect to the crust -- a motion of considerable interest to geodesists and geophysicists. Yet, the fundamental definition of the CIO is based on optical observations whose precision does not compare with that available currently from modern techniques. Indeed, the fundamental definition of the CIO is no longer achievable since not all of the optical instruments used in the definition are in service any longer. The BIH zero of longitude similarly serves as the origin for the measurement of geodetic astronomic longitude. Yet it has no acceptable definition and its accessibility is in question. The demonstrated fact of crustal motion must be included in future definitions of the terrestrial system, and it is likely that a set of geodetic observatories will be required to monitor this motion.

What will all of these improvements permit us to accomplish in the next few decades? Certainly crustal motion models will be used commonly and it is likely that the forces driving these motions will come to be understood. Relativistic effects in the definition of coordinate systems will be validated and incorporated.

The improved accuracy with which we will be able to obtain Earth orientation information will permit us to reach a new understanding of polar motion. It will be possible to relate geophysical phenomena to the excitation of polar motion and answer the questions regarding the role of earthquakes in exciting polar motion. Also, the improving ability to measure the atmospheric angular momentum will allow us to determine the role of the atmosphere in the motion of the rotational pole. Within the next few decades it is entirely possible that the enigmatic Chandler motion will be understood for the first time.

Currently, controversial questions regarding the variations in the Earth's rotational speed will also be answered largely through the improvement in observational accuracy. By increasing the time span over which we can obtain estimates of the excess length of day with unprecedented accuracy we may even be able to establish the nature of the tidal deceleration of the Earth's rotation. Improved astronomical observational information will be forthcoming regarding the "decade fluctuations" and with commensurate improvements in monitoring the magnetic field of the Earth we might be able to establish the nature of the coupling between the core and the mantle. Since the suspected relationship between the variations in the length of day and those of the westward drift of the non-dipole field currently provide the only observational information on core-mantle coupling, this will lead to a very important improvement in our understanding of the Earth. Just as with polar motion, the improvement in observational accuracy will enable us to understand the impact of various geophysical phenomena on the rotational speed of the earth.

Along with this improvement in observational accuracy we might also expect the possibility of an improvement in the time resolution and the timeliness of the Earth orientation data. We will establish the frequency with which these observations must be made to satisfy user requirements and reduce the time between the actual observation and reduction of the data. This fact along with our improved theoretical understanding will improve the accuracy of Earth orientation predictions which, in turn, may allow us to improve the accuracy of real-time positioning on the surface of the Earth using geodetic satellites.

In summary, it appears that the next few decades will indeed be an exciting time in the area of reference systems and Earth orientation.

THE ROTATION OF THE EARTH

G.A. Wilkins
Royal Greenwich Observatory,
Herstmonceux, Sussex, England

1. What are the causes of the secular variation in the rate of rotation and how have their effects varied since the formation of the Earth?
2. What are the causes of the decade fluctuations in the rate of rotation and what can we learn about the interior of the Earth from studies of these fluctuations?
3. What are the causes of the short-period variations in the rate of rotation and can we utilize our knowledge about these causes to predict the departure of universal time from atomic time?
4. How best can we use improved determinations and theories of the precession and nutation of the axis of rotation to give us new information about the interior of the Earth?
5. What are the sources of excitation and damping of the motion of the axis of rotation within the Earth and what are the causes and effects of the secular motion of the axis?
6. What are the relationships between the long-term variations in the rotation and climate of the Earth?

EARTH ROTATION

E.R. Mustel
Soviet Geophysical Committee,
Molodezhnaya 3, Moscow, USSR

To find a mechanism of generation of the Chandler motion of the Earth's rotation axis it is necessary to study the deformation of masses inside the Earth which causes the shift of the main inertial axes, migration of points over the Earth's surface and also variations in the gravity acceleration of the Earth's surface: the main problem is to find the energy source for these deformations and to locate the area of dissipation of this energy.

GRAVITY FIELD OF THE EARTH AND THE DYNAMICS OF ITS INTERIOR

S.K. Runcorn
School of Physics, University of Newcastle
Newcastle-upon-Tyne, England

Vening Meinesz was alone among geophysicists of his time in seeking a dynamic model of the Earth's mantle: seismologists and geodesists alike sought to derive from their data accurate models of its constitution and structure. His development of measurements of gravity at sea settled the controversy whether the chemical constitutions of ocean floor and continents were different, or whether the ocean floor were collapsed continents. His discovery of the negative gravity anomalies over the ocean trenches was the first geophysical evidence that raised the question of mantle convection. His chief argument, however, now forgotten, concerned a result he obtained from Prey's spherical harmonic analysis of the ocean-continent distribution; that the dominant terms, apart from the first, were degrees $n = 3, 4$ and 5. From the marginal stability theory of convection he showed that the same harmonics would be the ones to develop in convection in the Earth's mantle. This explains the paradox that Vening Meinesz, in spite of being such an advocate of mantle convection, remained until his very last year a strong opponent of what would now be adduced as the most compelling evidence for convection: continental drift.

Vening Meinesz and Heiskanen took as the basic axiom of geodesy that there exist no long wavelength departures of the geoid from the isostatic model, which for him was a consequence of the fluid behavior of the mantle. So again we have a paradox that the discovery of such low degree non-hydrostatic terms in the geoid was made in pursuit of evidence for the rigidity of the mantle.

Jeffreys' analysis of about 2000 measurements of g over the land and sea and his first map of gravity anomalies, up to those of 3rd degree, showed the now familiar lows over the Caribbean and the Indian Ocean and highs over the West Pacific and North Atlantic; but it was not widely accepted at the time because it was thought that the use of spherical harmonic analysis on data so poorly distributed could yield spurious results. The first satellite geoids confirmed the existence of these anomalies, though in the Pacific there was no agreement with Jeffreys' map. Since then the satellite determinations of the geoid have provided one of the greatest ever accessions of fundamental data on the Earth's interior.

Many followed Jeffreys in supposing that the density differences in the mantle implied by the geoid were retained by the finite strength of the mantle: their origin in primeval processes of accretion or differentiation. Various models of random distribution of sources were popular, especially following Kaula's rule where the strength of the harmonics decreased with degree.

However Runcorn, concerned with reconciling this new discovery and the evidence for continental drift, suggested that the geoid arose from the density differences on horizontal surfaces driving mantle convection. He showed that the Navier-Stokes equation of the mantle simplifies to

$$\mu \nabla^2 \underline{v} = -\Delta p + g \Delta \rho$$

as the other terms are negligible. Taking $\mu = 10^{21}$ poise and velocities somewhat greater than the plate motions $\Delta \rho = 10^{-5}$. Thus convection is quantitatively adequate to explain the geoid.

This explanation was not generally accepted for many years, partly because of erroneous ideas about the mechanical properties of the mantle, especially that the viscosity of the lower mantle was so high as to preclude convection (and therefore could consequently contain sources) and partly because the false idea that general mantle convection rose below the ridges and sunk at the trenches suggested a close correlation between the highs and lows of the geoid and plate tectonics. However, the extensional phenomena at the ridges was explained by the movement apart of the plates and features of the ridges precluded them as loci of the uprising general convection. Therefore, while opinion has solidified that the geoid arises from convection in the mantle, attempts to understand the relationship to the present plate motions has progressed slowly.

It was early recognized that the distortion of the surface in a convecting shell contributed to and could reverse the sign of the gravitational anomaly arising from the negative density in the rising column and the positive density anomaly in the descending current. For the case of the Moon it was indeed proved that the positive gravity anomaly was over the rising current and conjectures that the opposite might be true in the Earth were made, for the simple assumption of constant viscosity and gravitational acceleration proportional to radius (appropriate for the Moon) did not hold for the Earth.

Recently seismologists have made a great breakthrough in studying travel time anomalies, showing that velocity differences of over 1% are found in a lower and upper mantle over horizontal surfaces. These lateral variations are different for the upper and lower mantle, while showing respectively some correlation with plate tectonics and with the geoid. It seems difficult to maintain that these seismic anomalies are simply density, as three orders of magnitude separate them: attempts to arrange that distortion of the boundary almost exactly cancel the gravity anomalies arising from such density differences, inferred to exist in the deep mantle, seem insecure and it seems more likely that solid state creep involving the growth of new crystals in the stress field of convection results in mantle wide anisotropy: differences in velocity along and perpendicular to crystallography axes are of the order of a few percent.

GEOID INTERPRETATION

P. Biró
Department of Geodesy
Technical University of Budapest
Hungary

What is the true physical interpretation of the geoidal undulations?

Geodesy provides a quite detailed representation of the geoid as an equipotential of the Earth's gravity field. Geoidal undulations refer to a conventional geodetic reference system especially to the reference ellipsoid being an equipotential of the normal gravity field. Geoidal undulations determined by geodetic methods are generated by "disturbing masses" i.e. by the anomalies of the mass distribution of the real Earth with respect to that of the reference body. But using the theory of the level-ellipsoid for the determination of the reference system the mass distribution of the reference body remains unknown.

The problem to be solved: what possible distributions of the real Earth's masses can be found starting from geoidal undulations and how to do it? (The same problem can be started from deflections of the vertical or generally from the disturbing potential.)

THE STRUCTURE AND DYNAMICS
OF THE EARTH'S DEEP INTERIOR

T. Nagata and S. Akimoto
The University of Tokyo, Tokyo, Japan

A. Introduction

Theoretical studies on the most probable process of the condensation and accretion of the primordial solar nebula and a resultant formation of the Earth have been gradually developed on the basis of the knowledge of the solar chemical composition as well as the petrographical and chemical composition of the meteorites. However, our present knowledge of the structure and dynamics of the Earth's core and mantle is still far from real precise-scientific understanding. The main difficulty arises from the absence of basic fundamental data of physical properties of materials constituting the Earth's deep interior under high pressures up to about 4 Mbar and under high temperatures up to about 5000°C. There are several fundamental problems to be solved with a well-balanced combination of the experimental high-pressure/high-temperature physics of the materials of the Earth's mantle and core, the quantum-mechanical understanding of these materials *in situ*, and the celestial-mechanical establishment of the macroscopic model of the constitution and dynamics of the Earth's interior. Suggested individual problems would be as follows.

B. Key Questions

1. How did the Earth's core form?

This question could be solved through experimental investigation of high-pressure/high-temperature behavior of the primordial material of terrestrial planets. As a first step, high pressure/high-temperature phase equilibrium study among Fe-Ni metal and silicates will provide basic information. Since it has already been demonstrated that presence of a light element such as H in the Fe-Ni dominant core markedly lowers the melting temperature of Fe-Ni alloy, comprehensive examinations of the phase relations among Fe-Ni and H, O, S under high pressure will be primarily significant. The core temperature could also be reasonably evaluated by experimental observation of the melting curve of the core material under the required pressure. The shock compression experiments would realize the necessary pressures of 1.4 to 3.3 Mbar.

2. How did the Earth's mantle become layered?

By analogy with the evolution of the moon, it has recently been suggested that the Earth may have had a magma ocean at its accretion period. The presence of the magma ocean would have had profound implications for the primary stratification of the mantle into upper mantle, transition zone, and lower mantle. The origin of seismic discontinuity separating transition zone and lower mantle could be identified to a chemical composition boundary or to a phase boundary, through the phase relation research of the mantle materials under high pressure around 250 kbar. Precise knowledge of the high-pressure melting characteristics of the mantle materials is particularly important.

3. How much amount of H_2O is reserved in the Earth's deep interior?

   The distribution of H_2O within the Earth's deep interior is still one of unsolved problems with respect to our H_2O-rich planet. It has recently been demonstrated that a silicate in the system MgO-SiO_2-H_2O can stably exist on the condition of 250 kbar and 2000°C. A considerable reduction of the melting temperature of mantle rocks is also expected in the presence of H_2O. More extensive studies on hydrous silicates under very high pressure would be required.

4. What are the vertical and horizontal scale of mantle convection?

   The concept of a subduction of lithosphere into the mantle is now widely accepted, but little is known about possible change of the subsiding lithospheric materials in the mantle. This problem is related to the lateral inhomogeneity of the mantle revealed by recent Earth-tomographical research, and to occurrence-mechanisms of the deep-focus earthquake. The mantle convection hypothesis also may have to be examined on the basis of material changes in the mantle conditions. Accurate knowledge on the rheological properties of high-pressure minerals is likely to provide important constraints for the mantle convection.

5. How did the planets evolve?

   Comprehensive understanding of the structure and processes of the terrestrial planets and their satellites in comparison with the Earth is one of urgent problems in geophysics. The high-pressure metamorphism and the impact metamorphism are to be the common base for comparative planetology. The structure of Jovian planets could also be elucidated by investigating behavior of hydrogen and helium under very high pressures of 2 to 5 Mbar, which may not be impossible to be produced in the near future. Present directions include the synthesis and characterization of metallic hydrogen and measurements of equations of state of hydrogen-helium mixtures.

THE CORE OF THE EARTH

C.G.A. Harrison
Rosenstiel School of Marine & Atmospheric Science
University of Miami
Florida, USA

Our knowledge of the core of the Earth comes from seismology, from a study of the Earth's magnetic field, and from Earth rotation. Significant advances have been made in all of these three disciplines recently, and consequently our knowledge of the core continues to improve rapidly. It is to be hoped that in the next few decades, some major questions about the core will have been answered. Some of these questions are: What are the surface patterns of core motion and how rapidly do they change? What is the interaction between the core and the mantle? What is the source of heat which keeps the geodynamo powered, radioactive decay, latent heat of freezing, or gravitational energy? Is the core cooling down? What are the deeper patterns of core convection? What type of dynamo is responsible for the geomagnetic field? Why is there such a large change in the rate of geomagnetic reversals through geological time? What is the nature of the thermal link between core and mantle, and are mantle processes important in controlling the geodynamo?

The advances in our observational capability consist of satellite measurements of the field and space measurements of Earth rotation, seismic tomography, more complete paleomagnetic data sets and theoretical advances in determining core surface motions from secular variation studies. These are briefly discussed below.

1. Satellite Observations of the Earth's Magnetic Field.

   POGO and MAGSAT observations have greatly enhanced our knowledge of the Earth's magnetic field, especially at the longer wavelengths where the field of the core dominates that from the crustal rocks. As yet, we do not know as much about the changes in the field, the secular variation, as we know about the magnitude of the field components. But with new planned satellites this deficiency should soon start to be solved. The Geopotential Research Mission (GRM) will provide another snapshot of the field, but even more accurate than that produced by MAGSAT because of its lower altitude and because some of the vector orientation problems suffered by MAGSAT will hopefully not occur with GRM. The Magnetic Field Explorer mission(MFE) along with a companion program by France (Magnolia) will measure the changes in the field over a number of years from an altitude higher than MAGSAT. These space observations should then give us an idea of the time varying field over periods from one month to twenty years (with some unfortunate gaps). In order for the fullest use to be made of the satellite data it is important that magnetic observatories be operational, and if possible, their numbers enhanced during the period of satellite data acquisition. This will allow for better removal of the external field.

2. Paleomagnetic Observations

   Observations over longer time scales, using paleomagnetic techniques such as measurements on lake sediments, are now producing data which can enhance our understanding of the secular variation over periods up to 10^4 years. In particular, some researchers are convinced that periodicities are seen in some patterns of secular change. This implies a memory within the core which is much longer than had been thought possible.

3. Theory

   Theoretical advances have been made in determining core surface motions from observations of the surface magnetic field. The old concept of a fairly uniform longitudinal drift has been shown to be too simple-minded. Although much theoretical work remains to be done, it is probable that fairly complete information about core surface motion will be available at longer wavelengths within the next decade, provided that the observational program outlined above is carried out. The results of this sort of calculation will be invaluable for dynamo theorists.

4. Earth Rotation

Observations of the variations in the length of the day by space techniques have allowed this to be much more precisely determined. The other two components of Earth rotation, namely the position of the instantaneous spin axis, have also been better determined. It is possible that some of the intermediate period signal in these observations may be caused by core-mantle interactions. This again will add important information to our core data set.

5. Seismology

It is to be hoped that seismology will aid us in understanding core dynamics. Recent advances in seismology have allowed the mantle to be treated as a medium which has lateral variation as well as the more obvious radial variation. If such work can be continued into the core it might aid us in understanding the gross features of core motion, such as the major convection patterns which exist below the surface of the core, which cannot be directly inferred from magnetic field measurements.

MANTLE STRUCTURE

V.P. Trubizin
Soviet Geophysical Committee
Molodezhnaya, Moscow, USSR

1. What is the nature of the density discontinuity near the depth about 670 km: contribution of chemical change and phase transition?
2. Estimation of average viscosity of lower mantle, more accurate, than up to the order of magnitude. New kinds of data for such estimation.
3. What is the shape of convection cells in the mantle, in particular -- their vertical dimensions? Are their horizontal dimensions commensurable with the oceanic plates?
4. What is the contribution of gravitational differentiation into energy balance within the Earth, in particular -- into the present heat flow?
5. What is the depth-dependence of chemical composition, and in particular of radioactivity, within the mantle? What happens near the bottom of the mantle: disintegration into oxide or formation of the new complicated composite?
6. Present content of light elements and in the mantle.

MANTLE DYNAMICS

P.N. Kropotkin
Soviet Geophysical Committee
Molodezhnaya 3, Moscow, USSR

1. To find a model of the motion of the matter in the mantle, which would explain the observed stress-pattern in the lithosphere. In particular, how to explain a wide occurrence of large horizontal stresses. The existence of the monolithic, rigid, cold roots of the cratons down to the depth of 400-600 km below the Pre-cambrian shields. These roots should move together with the plates; that is difficult to reconcile with the plate tectonics. I expect that the hypothesis of the pulsating Earth will be essential to solve these problems.
2. To find a model, which would explain the correlation of temporal variations of seismicity with the following phenomena: quasiperiodical (5-20 years) change of the velocity of rotation of the Earth; amplitude and transient period of Chandler wobble; intensity and velocity of the westward drift of geomagnetic field, etc. I expect that the variation of the radius of the Earth is a common reason of these correlations.

MANTLE CONVECTION

A.S. Monin
Soviet Geophysical Committee
Molodezhnaya 3, Moscow, USSR

Measurements of the mantle convection velocity field (10^0-10^1cm/yr) could make it possible to reconstruct configuration of the mantle convection cells and thereby to finalize the construction of theoretical and factual bases of lithospheric plate tectonics which is a matter of principle for the Earth sciences on the whole and for planetology in general.

The contemporary configuration is supposed to be bi-cellular of a tennis-ball pieces type (scarcely strictly symmetrical) with a rise "equator" along the seam and subsidence "poles" in the middle of the pieces.

The key questions are:

1. Does this configuration determine the geoid that is the sea level measured from satellites after the correction for the effects of wind, internal currents and temperature?
2. Does this configuration determine the upper and lower boundaries of the asthenosphere (these boundaries of the asthenosphere can be outlined from electromagnetic or seismological data)?
3. Can this configuration be reconstructed from the velocity field of continental motion (data on velocity field can be obtained mainly by satellite aided technologies, e.g. laser ranging)?

CRUST AND MANTLE

I.Kh. Khamrabaev
Soviet Geophysical Committee
Molodezhnaya 3, Moscow, USSR

Three key problems are:

1. To clarify the correlation mechanism between the Earth's mantle and crustal peculiarities and magmatism and metallogeny.

2. To find out why no earthquake sources are located in granulite-basite ("basalt") layer.
3. To find out the causes and rates of the Earth's crust growth in various regions of the Earth.

LITHOSPHERE DYNAMICS

A.M. Gabrielov
Institute of Physics of the Earth
USSR Academy of Sciences, Moscow

1. What is the physical mechanism and mathematical model of the time delay in the earthquakes sequence like, i.e. in redistribution of stress after each earthquake? This time-delay, in particular, determines the time scale of post-earthquake activation such as aftershocks.

    The earthquakes are known to appear in time-space clusters. Most of these clusters are presented by a main shock and its aftershocks. It is worth mentioning also double earthquakes and interrelated strong earthquakes in adjacent seismic regions. At first glance it seems easy to explain this clustering by stress redistribution after each earthquake: stress exceeds strength at new places, and new fractures appear. However, with elastic stress redistribution and brittle fracturing adopted in many models, this process goes with the velocity of elastic waves, i.e. almost instantly. At the same time real aftershock sequences continue from several days to several years. Therefore for adequate description of seismotectonic process it is necessary to introduce some time-delay in stress redistribution after an earthquake and/or delay of fracturing when strength is exceeded.

    Several mechanisms of this time-delay can be suggested such as fatigue, viscosity, diffusion of fluids, stress corrosion, phase and petrochemical transitions. However, qualitative similarity of the process of clustering in regions with different tectonic structures allows to suggest some general property of self-organization of large non-linear systems for explanation of this process.

    An adequate model of clustering of earthquakes must explain some known features of aftershock sequences such as energy-frequency law, Omori's law, spreading of the aftershock area with time, etc. It has also to reveal the parameters of the medium that influence the level and duration of the post-earthquake activation. Such a model is necessary, in particular, in the problem of earthquake prediction to understand the nature of the "burst of aftershocks" precursor, i.e. anomalous number of aftershocks of some intermediate earthquakes in the period of preparation of a strong earthquake. It is now the only precursor with statistically proved significance. At the same time there is no adequate physical model of this phenomenon.

2. What is the physical mechanism and mathematical model of the re-establishing of strength ("healing") after an earthquake like?

    It is known that earthquakes happen at the fault zones presented by a hierarchy of volumes (blocks) divided by boundary layers (faults). An earthquake is usually associated with discontinuous deformation and fracturing of rocks. Here arises a fundamental question: why the fault zones have not been absolutely crushed by nowadays. In particular, how the earthquakes may reoccur at the same places.

    Different hypotheses were suggested to answer this question:
    a) Fault sides slip without destruction. The strength is re-established due to friction.
    b) The healing occurs due to physico-chemical processes. (It should be mentioned that the characteristic time of these processes is ~10000 years.)
    c) During the earthquake process there is no destruction, just recombination of the small blocks forming the fault zone.

    None of these explanations are proved, and they do not predict new phenomena.

GEODYNAMICS OF SUBDUCTION

Seiya Uyeda
Earthquake Research Institute
University of Tokyo, Tokyo, Japan

Introduction

In plate tectonics, subduction of oceanic lithosphere is the basic model for the tectonics of trench-arc-backarc (denoted as T-A-BA) systems. However, the subduction models cannot readily explain some of the major characteristics of T-A-BA systems. There are four basic interrelated unsolved problems.

Unsolved Problems

1. Why are arcs in arcuate shape but not always?
2. Why is the upper mantle "hot" under the inner zone of the arc? (Origin of arc magmatism and high heat flow.)
3. Why is the stress "tensional" in the backarc areas? (Origin of backarcs basins.)
4. Why do the arcs rise? (Origin of arc mountain belts.)

    To understand these problems, it is suggested that solution of the following more specific problems might be helpful.
    a) What is the role of the sphericity of the Earth?
    b) Origin of the trench outer swell and its diversity.
    c) Shallow outer-arc seismicity and its diversity.
    d) Real nature of the Wadati-Benioff zone.
    e) What determines the magnitude of giant inter-plate earthquakes or the degree of mechanical coupling between upper and lower plates?
    f) Time-space distribution of arc volcanic and plutonic rocks.
    g) Source material and depth of arc magma production.
    h) Does descending oceanic crust melt?
    i) Possible role of water and CO_2 in magma production.
    j) Is "frictional heating" important in thermal process of subduction?

k) Thermal process of anataxis.
l) Possible effects of higher subduction rate.
m) What determines the occurrence of accretion and erosion?
n) True nature of major decollement.
o) Why some arcs develop backarc basins and others do not.
p) What does a collision of buoyant-features and trench do on T-A-BA systems?
q) What is the role of strike slip faulting?
r) Does major mountain building require collision/accretion or not always?
s) Relative importance of lateral compression and pluton upheaval in mountain building.
t) Is crustal shortening at collision zones realized by doubling or by accordian-like thickening?

State of Art

Considerable advance has been made so that at least partial solutions to some of the above listed problems have been found. Especially, the problems related with "diversity" have been extensively studied by what the present author calls "comparative subductology". Many aspects of diversity are now interpreted in terms of diversity in the degree of mechanical coupling between overriding and subducting plates. However, the mechanism(s) controlling the degree of mechanical coupling is still a matter of debates. Problems related to collision and accretion have been proven to be of fundamental importance. But does this mean that the Andean type orogeny is ruled out?

Generally, it can be said that as far as seismic activities are concerned, the new global tectonics have been quite successful, although there still remain some major problems, such as the diversity in seismicity for different arcs, the physical mechanism of deep earthquakes, and their double zone characteristics and the maximum depth of penetration of subducting slabs.

As to the thermal features of subduction zones, the situation is different because subduction of cold oceanic lithosphere intuitively does not seem to result in either high heat flow in backarc regions or arc magmatism. Low Pn velocity and low Q under the arc and backarc regions and the existence of the aseismic front also testify to the anomalously high temperature in the mantle wedge above the slab. This is the problem (2) mentioned in the Introduction. Closely related to this is of course the problem (3), namely the origin of backarc basins. Some backarc basins are believed to have been generated by backarc spreading and dated magnetic lineations in many backarc basins testify to its validity. But why does the backarc spreading process ever take place behind arcs where oceanic lithosphere is thrusting down? As to the problem (4), namely the origin of arc mountain belts, since the advent of plate tectonics, subduction has been assumed to cause mountain building. The Cordilleran (Andean) or the Pacific-type orogeny has been distinguished from the Alpine-Himalayan orogeny which requires continental collisions. However, since the late 1970's, so called accretion tectonics have been postulated as the basic process for mountain building. It is said that numerous buoyant features on the ocean floor are accreted to continents during subduction and become mountains, and simple subduction of smooth oceanic plate cannot give rise to sufficient deformations for building major mountain ranges. Key to this question may be whether or not the Andean mountains consist of accreted terranes. At present, it is an open question. Thermal process associated with collision/accretion and its role in mountain building is far from any solution.

The problem (1) of why orogenic zones, in particular, the subduction zones, have circular or arcuate shape jointed by cusps has been discussed from time to time but, in spite of its obvious importance, conclusive solution has not yet been reached. As soon as the advent of the concept of plate subduction, it was pointed out that a subducting spherical lithospheric plate, if flexible but inextensible, may be bent inwards through an angle α on and only on a circle whose radius of curvature is $\alpha/2$. It was found, however, that this curvature-dip relation holds in some cases but not always. On the other hand, it appears also true that cusps between arcs are well developed where buoyant features are indenting the overriding plate. Thus, the competing two theories on arc-cusp formation at subduction zones are at present almost equally persuasive.

As stated in the Introduction, although much progress has been made in understanding the origin of T-A-BA systems and subduction tectonics, we are still far from the ultimate solutions of a number of fundamental problems. Solving these problems is of paramount importance not only in understanding the fundamental tectonics of the Earth but also in realizing a workable prediction system on hazardous earthquakes and volcanic eruptions.

THE RHEOLOGY OF THE LITHOSPHERE

D.H. Matthews, F.R.S.
BIRPS, University of Cambridge, England

The theory of Plate Tectonics, widely accepted by 1970, makes a simple assumption about the rheology of the upper part of the oceanic lithosphere: it's rigid, within plates. Plates can be recognized by the seismicity at their edges. As a first approximation the theory has been outstandingly successful and its numerical predictions apply very well -- but only within the oceans. On the continents Plate Tectonics doesn't work at all; you cannot recognize aseismic plates within a framework of seismic plate boundaries on land. Most geophysicists never thought that it would work there.

In the last fifteen years we have been gathering data about the ancient, scarred, lithosphere under the continents. Geological observations have been made over almost all the land surface of the Earth. They can be made, using drillholes or mountains, over a vertical distance of a few kilometers; their extrapolation, by a factor of ten, down to the base of the lower crust, the Moho, is dubious, to say the least of it. We get clues from xenoliths and from metamorphic rocks exposed at the surface, but the physical state of rocks in the lower crust and upper mantle is a matter for argument, and even their mineralogy is in doubt. Do they have fluids with them, or not? Is the heat transported entirely by conduction, or by conduction and convection? How is electricity conducted?

Wide angle seismic refraction and reflection, earthquake mechanisms, V_p/V_s, magnetotelluric studies, and most recently, normal incidence seismic reflection, all enable statements to be made about the physical properties and structures of the rocks of the lower crust and upper mantle, the lithosphere. Seismic reflection has better resolution than the earlier techniques, but it isn't much use without a more detailed knowledge of the velocity structure than we presently have. And in the absence of detailed information about Vp/Vs, we are not able to tell the geologists what they want to know, "What are the rocks down there and do they have fluids in them?". These are the problems with which we are engaged and which, with luck, we shall solve to the satisfaction of most scientists within the next ten years.

The problem for the future is to understand the rheology of the continental lithosphere, to produce a mathematical model able to predict the response of the lithosphere to a given stress field. Present knowledge (Molnar & Tapponier; McKenzie & England; Meissner, Kusznir) suggests that the parameters that must be specified include temperature, fluid concentration, and the mineral composition of the rocks. The hope is that, given some knowledge of mineralogy and of structures presently observable in the crust, one would be able to calculate temperatures and fluid concentrations within the crust at the time of orogenesis (or, of course, any other combinations of variables). In turn this might lead to an understanding of the varying temperature profile during the period of basin formation and crustal thickness relaxation subsequent upon mountain building, and thus perhaps to a better understanding of hydrocarbon maturation, the movement of fluids during overthrusting and low angle normal faulting and the formation for quartz veins and ore bodies.

LITHOSPHERE DYNAMICS

I.P. Dobrovolsky
Institute of Physics of the Earth
USSR Academy of Sciences, Moscow

Two problems are:

1. Construction of a mathematical model describing the regular process of deformation of the Earth crust and lithosphere in a large region (Middle Asia, North America, etc.). Determinations should be made of the mechanical rock properties, averaged over hundreds and thousands of kilometers, and of the deformational behavior at the boundaries of the region, i.e., the boundary conditions.
2. Construction of theory for a cycle of a spatially isolated earthquake. The theory of the earthquake cycle unites the theory of preparation, the theory of the source and theory of aftershocks.

STRAIN AND STRESS OF THE EARTH CRUST

H. Kautzleben
Central Institute for Physics of the Earth
Academy of Sciences of the GDR

1. What is the distribution of strain and stress within tectonically active regions in comparison with that in a tectonically stable region?
2. What is the temporal development of strain and stress within such regions after an earthquake?
3. What type of stress-strain-relation prevails in zones of tectonic weakness?

Progress in the development of efficient methods for earthquake prediction depends to a high extent on investigations of strain and stress systems using relevant precise and highly operational methods of geodesy and geophysics.

ASTHENOLAYER OF THE CONTINENTAL CRUST, FACTS AND PROBLEMS

B.S. Volvovsky, I.S. Volvovsky, Yu.M. Sarkisov
Soviet Geophysical Committee
Molodezhnaya, Moscow, USSR

Does the common occurrence of serpentinised ultrabasites in the structures of various ages of continents and island arcs imply the presence of the ultrabasite layer in the crust itself?

The answer for the oceans in supplied by the new global tectonics.

As regards continents, another mechanism is more likely, i.e., the mechanism of hard phase penetration and concentration of the upper mantle ultrabasites in the middle part of the crust.

The deep seismic research implies that the crystalline part of the continental crust has a partially inverted stratification: the middle part of the section is composed of serpentinised subcrustal substratum transported there after the formation of the lower granulite-basite and the upper granite-gneiss crustal complexes. The formation of the mantle layer between the upper and lower crustal complexes is probably associated with the general extension of the old continental lithosphere and with the hydrostatic intrusion of the underlying asthenospheric material through a system of vertical channels to the upper lithospheric levels.

As a result of the long term of the process, during which the upper mantle rocks concentrated between the complexes, the latter were not only separated along the vertical but also laterally fragmented into rigid blocks of different size. Consequently, it seems more justified to characterize the section of the crystalline crust not as a packet of normally stratified layers but as a double-banded (continuous or broken) "brickwork" of rigid blocks cemented together by the plastic asthenolayer of mantle rocks.

This is confirmed first of all by the data of deep seismometry, i.e., the hatched character of the reflected waves field, the dynamic and kinematic peculiarities in the records of the major components of the field of refracted and converted waves, their mutual positions on the plane of the travel-time curve, the spatial regularities in the variations and interrelations of propagation velocities of longitudinal and shear seismic waves. This position of the asthenolayer is confirmed by the higher geothermal gradient at the depth of its bedding, lesser density, relatively low viscosity, low specific electrical resistivity.

The suggested hypothesis is important for applied

geology. It allows us to conclude that in fact the appearance of the new formation of plastic serpentinised layer under the granite-gneisses and its thermodynamic influence on the granite-gneisses is the general primary endogenous cause for the formation and development of all types of surface structures on continents. In the places where this influence was the greatest and where a complete rupture of the granite-gneiss complex occurred, the conditions were created for the intensive rise of serpentinised ultra-basites into the uppermost crust and for the formation of the eugeosynclinal subsidences above them. An essentially different process is typical when the mantle flow forming the serpentinite layer due to some reasons is reduced from a certain moment of time and the energy transmission of the layer itself becomes dispersed. In this case the serpentinites cannot pierce at any point the entire granite-gneiss layer causing differential vertical movements of its separate blocks and the formation of the platform and orogenic structures over them. This concept also deals with the problems of the fold-nappe deformations. The major cause of the latter is the same process of new formation of the sepentinite layer over which the granite-gneiss blocks move as on a slippery foundation in the horizontal direction; the blocks break off from their granulite-basite base and crumple and push in front and on the flanks the already existing basins creating new ones in their wake.

CRACKS IN THE CRUST: EVIDENCE AND IMPLICATIONS

S. Crampin
British Geological Survey
Edinburgh, Scotland UK

Shear-wave splitting, diagnostic of some form of effective anisotropy, is displayed along almost all ray paths in the crust. It is seen in the shear-wave window above small earthquakes; near the North Anatolian Fault in Turkey (Crampin et al. 1985); in Tedzhikistan, USSR (Crampin et al. 1986a); in several places in Japan (Kaneshima et al. 1986a, 1987); in Canada (Buchbinder 1985); in California (Peacock et al. 1987); and elsewhere. Shear-wave splitting is observed above acoustic events in Geothermal reservoirs (Kaneshima et al. 1986b; Roberts and Crampin 1986). It is observed by the hydrocarbon industry in shear-wave reflection surveys in 12 sedimentary basins across North America (Alford, 1986; Lynn et al. 1986; Willis et al. 1986), and in vertical- seismic-profiles (VSPs) in the Austin chalk in Texas (Johnson 1986; Becker and Perelberg 1986) and the Paris Basin (Crampin et al. 1986b). Splitting is also seen in VSPs in a variety of terrains in California (Majer and McEvilly 1985; Leary and Li 1986; Li and Leary 1986; Daley, Majer and McEvilly 1986). There are no contrary examples where suitable shear-waves in the crust do not display splitting, and some form of effective seismic anistropy is clearly present in most crustal rocks.

This shear-wave splitting is caused by propagation through the distributions of fluid-filled cracks and microcracks which are known to exist in crustal rocks (Crampin 1985a; Crampin and Atkinson 1985). The cracks are aligned by stress, by such processes as subcritical crack growth, into typically parallel vertical orientations, striking normal to the minimum horizontal stress direction. Such distributions of aligned cracks are effectively anisotropic to shear-waves (Crampin 1978, 1984) and are known as *extensive-dilatancy anisotropy* or *EDA* (Crampin et al. 1984; Crampin 1985a; Crampin and Atkinson 1985).

The recognition of aligned fluid-filled EDA microcracks along most ray paths in the crust is a fundamental advance in our understanding of the physical state of *in situ* rocks. The presence of EDA cracks is a unifying concept which explains and associates a number of previously inexplicable and unrelated phenomenon. Many of the complications of shear-waves can be interpreted and, in some cases, the first few cycles of the shear wavetrain can be modelled, for the first time, with synthetic seismograms. Since, cracks and stress are crucially important in tectonic and deformatory processes and whenever we drill, mine, or excavate, there are many applications to currently important problems: earthquake prediction (temporal variations in shear-wave splitting have been observed in seismic gaps, Peacock et al. 1987; Chen et al. 1987); estimating the internal structure of hydrocarbon and geothermal reservoirs; estimating the orientations of hydraulic fractures; estimating preferred directions of flow; among others.

This new understanding of the state of *in situ* rocks, and the ability to monitor and estimate some of the parameters of *in situ* cracks and *in situ* stress has important implications over a wide geological and industrial field. We are only just beginning to explore the implications and applications (almost all of the observations have been published only in the last two years), but it is already clear that these techniques allow us to estimate a whole new range of parameters describing the internal structure of the rockmass.

Seismic observations of *P*-waves are the basic tools for many Earth science investigations and applications. Shear-waves have hitherto been largely neglected, although it is easy to show that the shear wavetrain carries three or four times more information than the equivalent *P*-wavetrain (Crampin 1985b). Thus the ability to interpret and model shear-waves in terms of the internal crack and stress geometry in the rockmass is probably one of the most important advances in seismology for several decades. New parameters for specifying the interior of the Earth are available and a renaissance of Earth science activities can be anticipated.

References

Alford, R.M. 1986. Shear data in the presence of azimuthal anisotrophy: Diley, Texas, Expanded Abstracts, 56th Ann.Int.SEG Meeting, Houston, 1986, 476-479.

Becker, D.F. & Perelberg, A.I.,1986. Seismic detection of subsurface fractures, Expanded Abstracts, 56th Ann.Int. SEG Meeting, Houston, 1986, 466-468.

Buchbinder, G.G.R. 1985. Shear-wave splitting and anisotropy in the Charlevoix seismic zone, Quebec, Geophys.Res.Letters, 121, 425-428.

Chen, T.-C., Booth, D.C. & Crampin, S., 1987. Shear-wave polarizations near the North Anatolian Fault -- III. Observations of temporal changes. Geophys.J.R.Astr.Soc., in preparation.

Crampin, S., 1978. Seismic wave propagation through a cracked solid: polarization as a possible dilatancy diagnostic, Gephys.J.R.Astr.Soc., 53, 467-496.

Crampin, S., 1984. Effective elastic constants for wave propagation in anisotropic media, Geophys.J.R.Astr.Soc., 76, 135-145.

Crampin, S., 1985a. Evidence for aligned cracks in the Earth's crust, First Break 3, 12-15.

Crampin, S., 1985b. Evaluation of anistropy by shear-wave splitting, Geophysics, 50, 142-152.

Crampin, S. & Atkinson, B.K., 1985. Microcracks in the Earth's crust, First Break 3, 16-20.

Crampin, S., Evans, R., Ucer, S.B., et al., 1985. Analysis of records of local earthquakes: the Turkish Dilatancy Projects (TDP1 and TDP2), Geophys.J.R.Astr.Soc. 83, 1-92.

Crampin, S., Booth, D.C., Krasnova, M.A., Chesnokov, E.M., Maximov, A.B., Tarasov, N.T., 1986a. Shear-wave polarizations in the Peter the First Range indicating crack-induced anisotropy in a thrust-fault regime, Geophys.J.R.Astr.Soc., 84, 401-412.

Crampin S., Bush, I., Naville, C. & Taylor, D.B., 1986b. Estimating the internal structure of reservoirs with shear-wave VSPs. The Leading Edge, 5, 35-39.

Crampin, S., Evans, R. & Atkinson, B.K., 1984. Earthquake prediction: a new physical basis, Geophys.J.R.Astr.Soc., 76, 147-156.

Daley, T.M., Majer, E.L. & McEvilly, T.V., 1986. Analysis of shear-wave and P-wave data at the Salton Sea Scientific Drilling Program, Eos 67, 1116.

Johnston, D.H., 1986. Detection of fracture-induced velocity anisotropy, Expanded Abstracts, 56th Ann.Int.SEG Meeting, Houston, 1986, 464-466.

Kaneshima, S., Ito, H. & Mitsuhiko, S., 1986a. S-wave polarization anisotropy observed in the rift valley in Japan, Abstracts, Second Int.Work. Seismic Anisotropy, Moscow, 1986, 127.

Kaneshima, S., Ito, H. & Mitsuhiko, S., 1986b. Shear-wave splitting observed above small earthquakes in the geothermal area of Japan, Abstracts, Second Int. Work. Seismic Anisotropy, Moscow, 1986, 128.

Kaneshima S., Ando, M., & Crampin, S., 1987. Shear-wave splitting above small earthquakes in the Kinki District of Japan, Phys.Earth.Planet. Int., 45, 45-58.

Leary, P.C. & Li, Y.-G., 1986. VSP fracture study of Majave Desert hydrofracture borehole, Eos 67, 1116.

Li, Y.-G. & Leary, P.C. 1986. Seismic ray tracing of VSP in inhomogenous aligned fractured rock at Oroville, CA, Eos, 67, 1117.

Lynn, H.B. & Thomsen, L.A., 1986. Reflection shear-wave data along the principal axes of azimuthal anisotropy. Expanded Abstracts, 56th Ann.Int.SEG Meeting, Houston, 1986, 473-476.

Majer, E.L. & McEvilly, T.V., 1985. Fracture mapping using shear-wave vertical seismic profiling, Eos 66, 950.

Peacock, S., Crampin, S. & Fletcher, J.B., 1987. Shear-wave splitting in the Anza seismic gap, Southern California: temporal changes at one station -- a possible precursor, J.Geophys.Res., submitted.

Roberts, G. & Crampin, S., 1986. Shear-wave polarizations in a Hot-Dry-Rock geothermal reservoir: anisotropic effects of fractures, Int.J.Rock.Mech.Min.Sci. 23, 291-301.

Willis, H.A., Rethford, C.L. & Bielandski, E., 1986. Azimuthal anisotropy: occurrence and effect on shear-wave data quality, Expanded Abstracts, 56th Ann.Int. SEG Meeting, Houston, 1986, 479-481.

CONTINENTAL AND OCEANOGRAPHIC LITHOSPHERES

V.V. Beloussov
Soviet Geophysical Committee
of the Soviet Academy of Sciences Moscow, USSR

If we wish to achieve rapid progress in geosciences, the first thing we should do now is to get rid of the dogmatic fetters of plate tectonics. Everybody is aware that the plate tectonic fundamentalism has receded into the past. The plate tectonics idea was advanced as the reaction to the discoveries in the structure of the ocean bottom and, no doubt, its impact had a positive influence. This positive influence, however, was indirect rather than direct; it attracted a large group of specialists in exact sciences to the studies of the Earth and as a result serious achievements were made in the elaboration of new methods of research at great depths of the globe. In its turn this was followed by the accumulation of a vast amount of entirely new observation data whose value is intransient and independent of all theoretical conceptions. Anyway, the revolutionary effect of plate tectonics on the ideas in geosciences is obvious.

However, in the course of time, the fundamental strongholds of plate tectonics were steadily and rather quickly eroded and its field of application was gradually reduced. Only a few scientists now believe that this conception can be even partly applied to continents. This is a very serious circumstance and it deprives plate tectonics of its universal nature. Continents are much more complicated than the oceanic bottom by the composition of the material, by their structure, and by the history of their evolution. The period of their formation lasted four billion years, whereas the geologically documented history of the oceanic bottom does not exceed 170 million years. Does that mean that plate tectonics "works" only within a short period of geological time and only in those areas of the globe where the composition and structure of the lithosphere is to most monotonous?

But we should not get too excited over this restricted applicability of plate tectonics. It is highly improbable that this conception would still exist if such an essential element of it as subduction finally turns out to be a myth. Professor Uyeda, a well-known specialist in "subductology," indicated four basic problems of subduction as yet unsolved. Three of them are related to the subject matter of subduction: the thermal regime of the upper mantle, extension "behind" the subduction zone, the causes of island arcs uplift. The difficulties are not checked at that. The same author gives a list of twenty more issues which should be investigated in order to understand the process of subduction. Among them are the following:

a) Real nature of the Wadati-Benioff zone.
b) Source material and depth of arc magma production.
c) Is "frictional heating" important in thermal process of subduction?
d) Does major mountain building require collision/accretion, or not always?

If after twenty years during which the idea of subduction existed the enumerated problems remain unstudied and the answers after their study are as yet uncertain, then on what is the idea of subduction based? Let me remind you that deepwater drilling has as yet found no proof of subduction anywhere. It would be greatly straining a point to consider the results of drilling neutral anyway. It is more honest to accept them as negative.

We are witnessing now how plate tectonics becomes a certain verbal ritual obligatory to all who wish to belong to the "brotherhood." We are not only forced to look but we make ourselves search for any signs that would allow us to cry "Vive le Roi!", though the king is nude.

Plate tectonics is greatly compromised by unrestricted permissibility stemming from the absence of a mechanism of motion of plates, from the purely kinematic content of this conception, which means that any construction conceived by certain "rules of the game" accordingly becomes possible and fundamentally irrefutable. Plate tectonics lacks that quality of falsifiability which is the indispensable feature of scientific conception, as defined by the modern philosopher.

It is high time we acknowledge the truth. We should thank plate tectonics for the exciting minutes it accorded us and turn to current matters. First of all let us pause and take our bearings. Then recall everything that classical geology has accumulated in the course of two hundred years of its development about the structure, history and regularities of evolution of continents, all that plate tectonics has tried to trample down in the excessive rapture of its successes. The archives of geology contain bountiful information on the development of continental crust during billions of years and on its present-day life! And then let us try to apply to these geological data the new materials about the deep structure of the crust and upper mantle, which are obtained in the last decades partly owing to plate tectonics.

This confrontation will induce us to concentrate our attention on apparently two major problems:

a) heatmass transportation in the Earth, and
b) interrelations between the continental and oceanic lithospheres.

The first problem arises from the observed correlation between heat flow and other manifestations of warming of the crust and upper mantle, on the one hand, and the "excitation" of endogenic processes, tectonic, magmatic, metamorphic, on the other hand. The geological history of the continents shows that the areas covered by "excited" endogenic regimes (geosynclinal, orogenic, rifting, etc.) and quiescent regimes (platforms) and also the distribution of these and other regimes changed in the course of geological time. It is also essential that these changes reveal obvious demonstrations of both spatial and temporal regularity and a well expressed direction. Consequently, the heat flows from deep geospheres to the surface also changed their distribution and intensity with regularity. From this standpoint we may approach to the understanding of the pattern of heatmass transportation in the globe, establish the changes in the pattern which occurred during the geological time, and determine the spatial and temporal scales of this major process supplying the tectonosphere of the Earth with energy. I believe that this course of investigation, from observations to history and then to theory, is much more promising than the construction of speculative models of convection in the Earth's material, which hardly go beyond academic physico-mathematical exercises.

The second problem in a more strict definition is to find out the *historical* relationship between the two lithospheric types, continental and oceanic. Does continental crust appear in place of oceanic crust in the course of time, or, on the contrary, does oceanic crust somehow appear in place of continental crust, or do both these transformations occur in certain proportions. At the present moment the first assumption is the most popular. If nonetheless the appearance of the new oceanic crust in place of continental crust is admitted, then it is only in areas where continental crust is stretched and fractured.

The objective analysis of the problem brings us to the conclusion that in many instances the stretching of the crust cannot be regarded as the likely cause of thinning and splitting of the crust. Within continents there is simply no room for extension stretching and the structure of tectonic depressions, which are considered as a result of stretching, show no structural signs of this process. It is astonishing, however, what multitude of various events can have a simple explanation if we concede that in nature there is a process of *in situ* substitution of continental crust by oceanic! Moreover, the intermediate stages of the process are demonstrated in the thinning of the continental crust, also *in situ*.

This is not the occasion for discussions. As the initial step, let us face the results of that wonderful achievement, the deep drilling. The holes were apparently drilled with the implication that their data were comprehensively used. We should note, therefore, that the larger part of the holes in the oceans and seas, which reached the "basement," showed that the sedimentary section starts as a rule with shallow-sea sediments or even sediments of continental origin, and only later in different localities and time periods the shallow-sea and continental environments changed to deep sea conditions. It was revealed that in the Jurassic and locally also in the Early Cretaceous in place of the modern oceans there were mostly epicontinental shallow seas and locally dry land. The land and shallow seas in modern environment are underlayed by the crust of continental type. Should we believe then, that in the Jurassic and Early Cretaceous the land and shallow seas covered the thin ocean crust thus defying isostasy? If not, then where are those enormous blocks of continental crust?

To conclude, let us again formulate the major goals:

a) to generalize the data of classical historical continental geology;
b) to study the historic development of heatmass transportation in the Earth on the basis of distribution of the "excited" and quiescent regimes in the course of geological history;
c) to study all the aspects of relationships between the continental and oceanic lithospheres in terms of the possible substitution of the continental lithosphere by the oceanic *in situ* without its essential stretching.

SEISMOGENESIS, PREDICTION OF EARTHQUAKES, AND MITIGATION OF EARTHQUAKE LOSSES

F.F. Evison
Institute of Geophysics, Victoria University
Wellington, New Zealand

Objective

To increase understanding of the earthquake phenomenon so that time-variable earthquake hazard can be mapped on a regular basis as well as the average hazard already codified in seismic zoning maps, thus enabling a major advance to be made in the mitigation of earthquake losses in all affected countries.

Background

Knowledge of the earthquake phenomenon is at present sufficient for the accurate estimation of average hazard and the partial mitigation of earthquake losses through seismic zoning. This achievement relies on a large data-base comprising the locations and sizes of past earthquakes over long periods of time, together with information on the properties of the Earth as a medium.

Recent advances in knowledge need to be vigorously followed up in order to achieve the estimation of hazard for each coming period in particular, and to specify how this differs from the long-term average. So far, time-variable earthquake prediction has been undertaken only to a limited extent and with very limited success.

Problems

1. The physical processes of seismogenesis need to be further elucidated by all available means, including theoretical, laboratory and field research.
2. Systematic studies of earthquake precursors need to be carried out, building on the extensive existing literature on precursory phenomena of diverse kinds, with a view to determining to what extent occurrences of such phenomena are followed by earthquakes and to what extent occurrences of earthquakes are preceded by such phenomena.
3. The seismic gap model of major earthquake occurrence at plate boundaries needs further development and formal testing against the Poisson model.
4. The behavior of interplate and major intraplate faults as a function of time needs further study with a view to elucidating the asperity, barrier and other models.
5. The mitigation of earthquake losses through the estimation of time-variable hazard needs to be further studied, in relation to the existing means of mitigation through seismic zoning.

INTRAPLATE EARTHQUAKES

Long-Sheng Gao
Institute of Geophysics
State Seismological Bureau
Beijing, China

A. Introduction

It is concerned that large intraplate earthquakes should have some cause in addition to the local stress accumulation or stress concentration. The earthquake related phenomena, which can occur prior to, during or after the event, sometimes are pervasive rather than concentrated only in regions around the epicenter of the earthquake. Plenty of precursory phenomena have been thought useful for the purpose of predicting earthquakes. However, due to the fact that these phenomena are easily to be confused with phenomena of other origins, rules related to the preseismic phenomena are usually followed by uncertain or negative examples. Therefore, the basic mechanism of the seismogenic process needs further research.

B. Key Questions

1. What is the mechanism of stress transmission in the crust which passes the stress over a long distance through the huge and thin, in terms of the ratio of its thickness to extension, plate?

    A large earthquake only chooses a suitable place to occur after bypassing a long distance. This seems to depend not only on the stress level and the strength of rock mass but also other factors.
2. What is the mechanism of the coupling between the lower lithosphere and the upper crust which is considered as a genetic layer of intraplate earthquakes?

    A number of theories consider that the driving force of geodynamics comes not only from the horizontal direction but also from underneath, for example, from the convection within the mantle. If the dragging force reaches the bottom of the plate, how can this force pass through the lower crust, where it is considered as a rheological aseismic layer, to the earthquakes genetic upper crust?
3. What is the stress state in the crust, especially in the vicinity of the epicenter of the impending earthquake?

    The low stress environment which is deduced from the present available data requires an explanation of the mechanism that how the fracture and friction slip can occur in such a low stress environment.
4. What are the mechanical properties of water in the very high pressure and high temperature environment?

    A number of studies concern that the role of water is very important in the course of large earthquakes, though its existence and mechanical properties, especially the friction properties under HP-HT are only little understood through both theoretical and experimental approaches.
5. Should there be really certain precursors which can be

used for prediction before an impending large earthquake?

If someday we can daily plot the contours of stress or strain and all the important parameters reflecting the activities of the earth's crust, just like what meteorologists are doing on their weather maps, shall we be able to make deterministic forecast of earthquakes rather than probabilistic prediction, even the density of the observation network is dense enough?
6. What is the right model for earthquakes, in view of physics, mathematics, chemistry, petrology, and so on?

EARTHQUAKE RESEARCH

R.G. Garetsky, A.P. Emelyanov
Institute of Geochemistry and Geophysics
of the Belorussian Academy of Sciences

What is the contribution of tectonic activity of the plates (and old platforms) into seismicity -- both the global (general) and intraplates? This includes the following subproblems:

a. Development of a model of deformations of the plates including seismicity.
b. Genetic classification of intraplate faulting and fracturing.
c. Estimation of the influence of the global processes and the processes in adjacent active regions on the intraplate and on the dynamics of fluids and gases within the crust of the plates.

EARTHQUAKE RESEARCH

Keiiti Aki
Department of Geological Sciences
University of Southern California

1. What causes the upper limit of seismic frequency, so called f_{max}, observed in the acceleration spectra of the strong ground motion? Is it the effect due to seismic source, propagation path, or recording site condition?
2. What causes the apparent constancy of corner frequency independent of moment observed often for earthquakes smaller than a certain magnitude (usually around 3)? Is it the effect due to seismic source, propagation path or recording site condition? The observed constant corner frequency is sometimes close to f_{max}. Are there any causal relation between them?
3. If the slip-weakening type of friction law with a given critical weakening slip applies to all earthquakes in a fault zone, there must be a minimum earthquake for the zone. Is there such a minimum earthquake for a fault zone? If so, is the size of the minimum earthquake related to f_{max}?
4. The critical weakening slip appears to control various aspects of earthquake recurrence behavior in a fault zone, including the quiescence phenomena sometimes precursory to major earthquakes. Since the critical weakening slip is known to be about 1/1000 the size of cohesive zone which may be controlling f_{max}, there may be relation between the recurrence behavior of a fault and its f_{max} if f_{max} is due to source effect. Is there any such correlation?
5. If f_{max} is due to source effect, it may be related to the upper fractal limit of fault trace geometry or gouge particle distribution. Is there any such relation?
6. Does the earthquake fault rupture nucleate by the growth of a crack or by the failure of an asperity? Can the geological study of morphology of fault zone and the seismological study of earthquake rupture resolve this question? For example, irregular geometry of fault trace, such as branching, stepping, bending and intersecting can have a dual effect of favorable (due to stress concentration) and unfavorable (due to stronger fracture toughness) for nucleation.
7. What are the precursor times and success rates for various earthquake precursors such as (1) seismicity quiescence, (2) clustering, (3) Radon and other geochemical precursors, (4) water table, (5) tilt, (6) strain, (7) coda Q, (8) Vp/Vs, (9) source mechanism, (10) stress drop, etc. Reliable knowledge on them is desperately needed for an operational earthquake prediction.
8. The quality factor of the Earth's crust measured by the use of 1Hz coda waves of local earthquakes shows strong negative correlation with the occurrence of major historic earthquakes in China. Does the same correlation hold for other regions where historic records of earthquakes are available? If so, the long term seismicity can be reliably predicted for any parts of the Earth by a simple measurement on coda waves.

PHYSICS OF SOLID ROCKS

E.I. Shemiakin, M.B. Kurlenja, V.N. Oparin
Institute of Mining Sciences Siberian
Department of Academy of Sciences, USSR

Key Problems

1. To explain the zonal disintegration phenomenon of rocks around underground cavities. It is known that alternating rings of fractured and intact rocks are formed around underground cavities at the depth where the gravitational component of stress is near or higher than the rock strength compression limit. This is caused by rock fracturing along the direction of major stresses near the maximum of supporting pressure by fracturing and destruction of rocks in this area (i.e., in fact, discrete release of geomechanical energy). But the following questions remain obscure: the dynamics and kinematics of this process (the time factor), the peculiarities of its manifestations in different scales, and the possibility of its application in technology.
2. To construct a model explaining the sign-alternation of response of the rocks to the impact from explosions and, in this connection, a model of "self-loading" massives of rocks. The mechanism of the impact from explosions on rock massif is to a certain extent

analogical to the experiments on the fatigue of the materials with the sign alternating loading, which results in the displacements of different directions between blocks in the structural hierarchy of massives and in their oscillatory movements against one another. A spontaneous fatigue of rocks is discovered as the result of the removal of macrostresses (from fairly slow to explosive). These facts indicate that the model of "self-loading" massives corresponds to geomaterials with great accuracy. An essential role in this model belongs to the partial removal of macrostresses in the process of release of a certain rock volume from external stresses and their redistribution inside the volume in accordance with the blocks structure and topological peculiarities of rocks. These factors should be considered in the construction of mathematical and mechanical models.

3. To confirm experimentally the hypothesis of the possible existence of spontaneous rock fatigue "wave". The above mentioned facts about the "self-loading" of geomaterials show the possible existence of such a "wave", because the occurrence of cavities or weakened surfaces in massives with thermodynamical and geomechanical equilibrium causes the destruction of the massives. The problem is to find convincing experimental data to prove the existence of the "wave", its role in the redistribution of rock pressure in time and to describe this "wave" mathematically.

4. To confirm experimentally and theoretically the hypothesis concerning the existence of the "migration" waves from earthquakes. The existence of "migration fronts" of induced seismicity from strong earthquakes is known in seismology. These "migration fronts" spread with low velocities. A correlation is shown between the position of the fronts and the frequency of rock bursts. Several problems still remain unsolved: the propagation mechanism of low velocity waves in the crust; their analytical description; the possibility of elaboration of new methods of long-term prediction of seismic activity and of the frequency of rock bursts in certain regions; their connection with the hypothetical "waves" of spontaneous fatigue of massives.

MULTIPLE SOURCES AND MULTIGENESIS, THE PROBLEM FACING GEOCHEMISTRY, PETROLOGY AND METALLOGENY WITH SPECIAL REFERENCE ON SKARNS AND SKARN DEPOSITS IN CHINA

Tu Guangzhi
Institute of Geochemistry
Academia Sinica

With the recent achievements in Earth sciences, especially in the fields of isotope geochemistry, trace element geochemistry and experimental petrology, the traditional thinking that rocks and ores are products of a single-staged event with simple genesis and single source of rock-forming or ore-forming material no longer holds true. Multiple sources of material and multiple genesis of rocks and ores seem to be a general rule rather than an exception.

We now know that andesite can originate by at least five different ways. There are several genetic types of granitoids, each with its specific tectonic setting, source material and geochemical characteristics. Quartz veins with metallic sulfides can be brought about by either magmatic, metamorphic or atmospheric water.

In the western geological literature, the word skarn refers to a unique mineral assemblage of diopside, tremolite, garnet, epidote, actinolite and other Fe-Ca-Al silicates which occurs at the contact zone of granitoid and carbonate rock. Skarns with their various associated ore deposits have been thought to be the products of replacement between granitoid and carbonates caused by magmatic fluids.

We have distinguished in China five genetic types of skarns. The traditional contact-metamorphic skarns situating between granitoid and carbonate predominate. But there are other skarns. In the Lower Yangzi Area, skarns are observed at the contact between the Mesozoic intermediate-basic eruptive rocks and the sub-volcanics of similar composition which intruded the latter. These skarns are doubtlessly related not to granitoid, but to volcanic activities. A third genetic type of skarn is found in some migmatite areas in North China, Xinjiang and Liaoning where the skarn-formation is probably related to the process of migmatization. In certain cases, regional metamorphism can produce skarns. The tremolite-diopside layers, intercalated with normal sedimentary beds, are wide-spread in the Triassic of Fujieng and in the Devonian of Guangdong. These rocks owe their origin to hot-water sedimentary processes. No granitoids are found nearby.

Metals in the skarn deposits are generally thought to have been brought up by magmatic fluids. This may be the case in some skarn ores. But, we have found cases, where ore galena gives age of the country rock, rather than the intrusive. For example, in Laiyuan, Hebei, Cretaceous rhyolite porphyry intruded the Precambrian 1300 MA old carbonate, forming at the contact Fe-Zn skarn mineralization. Ore galena gives an age of 1300 MA by Pb-Pb method. It is interesting to note that some sphalerite in this deposit has abnormally high content of Mn. Besides, minor amount of Mn-sulfide is also found in the ore, which is very unusual for skarn deposits. Recent finding shows that the 1300 MA carbonate in this area is also Mn-rich. It is thus postulated that both Pb and Mn, and possibly Zn in the skarn ore may have been derived from the country rock, but not the intrusive.

Some of the Chinese skarn iron deposits offer another example of complexity of ore genesis. There has been vivid discussion as to the source of iron in some diorite-associated skarn ores. The Mesozoic diorite neighboring to the ore bodies has been bleached due to albitization. Furthermore, it is found that the size and content of the ore bodies is proportional to the intensity of albitization. It is postulated that the Middle Ordovician salt and gypsum layers which were intruded by the diorite stokes, probably furnish the Na for the formation of albite and the Cl for the mobilization of Fe from diorite. Fe was leached from diorite as Cl-complexes and redeposited as ore. Pyrite in the Fe-ore gives a $\sqrt{S^{34}}$ value of 15% in average indicating the source of S from Ordovician sulphate rather that from magma.

Therefore, as is the case with granites, andesites, quartz veins, there are skarn and skarn, skarn deposit and skarn deposit.

PHYSICAL VOLCANOLOGY

Lionel Wilson
Environmental Science Dept., Lancaster University, U.K.
and
Hawaii Institute of Geophysics, Honolulu

Comment: We need to persuade more geophysicists to become involved in the theoretical study of volcanic processes. In addition to the gathering of field data on all of the topics mentioned below, we are in great need of more theoretical modelling studies and laboratory simulation experiments in all aspects of volcanology.

Questions:
1. How can we obtain direct information on the factors controlling the formation of convecting eruption clouds, on the one hand, and fire fountains, pyroclastic flows and pyroclastic surges, on the other?
   [As yet, we have only rudimentary ideas on what fluid dynamic processes occur in the region from several vent diameters below to several vent diameters above the surface level in explosive eruptions. This is true of relatively steady discharges of gas and pyroclasts and of sudden, transient explosions. We need to develop field techniques, based on all forms of remote sensing, to obtain data from active vents -- especially those that involve any form of cyclic activity on a time scale of days to weeks, so that it is reasonably cost-effective to station people and equipment on the volcano continuously to await the next event.]
2. How can we obtain very detailed information about the near-surface, internal structure of volcanoes, especially to the extent that this controls -- or is controlled by -- the interplay between intrusive and extrusive events?
   [This need applies to the edifices of volcanoes producing all compositions of magma, from basaltic shields to strato-volcanoes, and includes mid-ocean-ridge centers and seamounts. Here again there is a need for much more instrumentation (for seismic, resistivity and deformation studies, for example). Also, we should be thinking about the engineering problems of producing easily transportable or even expendable equipment.]
3. How can we predict the shapes and lengths of lava flows in more detail than is currently possible?
   [This problem is intimately related to investigating the temperature distribution in a flow, and requires a much more detailed understanding of the rheological properties of volcanic fluids than is currently available.]

ELECTRICAL CONDUCTIVITY

R.J. Banks
Department of Environmental Science
University of Lancaster
Lancaster LA1 4YQ, U.K.

A. Introduction

The electrical conductivity of the mantle can be inferred from measurements of the electromagnetic response of the Earth to natural geomagnetic field variations. Other than seismology, this technique is the only one to provide much information about the vertical as well as lateral structure of the Earth. Although its vertical resolution is poor in comparison with that yielded by seismic methods, the information it offers is of great potential value, because of the dependence of electrical conductivity on temperature, on phase transitions, and the presence of melt. Unfortunately, in spite of considerable effort, our knowledge of the radial variation of conductivity is still poor.

B. Key Questions

1. What is the "background" conductivity of the uppermost mantle?
   The background conductivity at depths of 100-500 km is usually assumed to lie between 10^{-3} to 10^{-2} S m^{-1}, but such estimates are based more on laboratory experiments and guesses about the ambient conditions than on electromagnetic response data. A major problem is that we probably need to incorporate lateral heterogeneity into any realistic model.
2. Is there an electrical equivalent of the asthenosphere?
   If the asthenosphere contains partially molten material, an enhancement of the electrical conductivity would be expected and some models do show a conductivity "spike" in the depth range 100-400 km. Unfortunately, it is not clear whether such spikes are essential features of the models or whether they are artifacts of the inversion procedure.
3. What is the nature of the transition region?
   Most models show conductivity rising steeply in the depth range 400-1000 km to values of 1-2 S m^{-1}. The location of this "jump" and the nature of the transition is unknown. If the depth of the rise could be determined more precisely, it might be possible to link it to one of the seismic discontinuities in this region. Better knowledge of the conductivity might help resolve questions about the reality of whole mantle convection.
4. Does the conductivity decrease below 1000 km as some models seem to suggest?
   Modelling often assumes monotonically increasing conductivity. Many inverse methods generate models which involve both increases and decreases. How necessary are the decreases?

GEODYNAMO

F.H. Busse
University of Bayreuth, D 8580 Bayreuth
and Institute of Geophysics and Planetary Physics
University of California, Los Angeles

A. Introduction

While there appears to be general agreement among geophysicists that the geomagnetic field is generated by motions inside the liquid core of the Earth through the dynamo process, the particular mechanism of the realized geodynamo is not well understood. More detailed theoretical modelling and more accurate data on the geomagnetic field and its variation in time such as those provided by the MAGSAT mission are needed before the magnetohydrodynamic dynamo problem for the Earth's core can be elucidated. The problem of the geodynamo is a key problem first because of its own fundamental character and, secondly, because of its connection with several other basic geophysical problems as outlined below.

B. Key Questions

1. What is the strength of the toroidal component of magnetic field inside the Earth's core?

    Current estimates range from 10^{-3} to 10^{-1} tesla: Subtle effects such as the small, but finite electrical conductivity of the Earth's mantle or constraints following from observations of the geomagnetic secular variations must be used to infer information about the toroidal magnetic field. From the theoretical point of view the dependence of the toroidal field strength on the given physical parameters will be of primary interest.

2. What is the mechanism of geomagnetic reversals?

    Several theoretical ideas exist about the mechanism of the reversal of the Earth's magnetic field. But without a reasonably realistic model of the Earth's dynamo it is difficult to arrive at a convincing model of reversals. On the other hand, the increasing information about the variation of the geomagnetic field in the course of a reversal is likely to provide important constraints for models of the geodynamo.

3. What is the mechanism of electromagnetic coupling between core and mantle of the Earth?

    This question is motivated by the apparent correlation between changes in the length of the day and certain properties of the geomagnetic secular variation. Because of the unusual time delay in this correlation, the answer to the question is not straightforward.

4. What is the nature of the thermal coupling between core and mantle?

    Since paleomagnetic data suggest a connection between reversal frequency and tectonic activity, a coupling between the geodynamo and the activity of mantle convection may have observable consequences. The lateral inhomogeneity of the thermal boundary layer at the bottom of the Earth's mantle undoubtably induces motions in the liquid core. Thus spatial as well as temporal variations of the heat flux from core to mantle can give rise to correlations between properties of geomagnetic field and of the Earth's mantle.

5. What are common and what are distinct properties of planetary dynamos?

    The form of a dipole nearly aligned with the axis of rotation for nearly all planetary magnetic fields suggests strong similarities of the dynamos operating in the electrically conducting cores of the planets. On the other hand, the energy source driving motions in the core differs for various planets and the recent Voyager observations of the magnetic field of Uranus have revealed an unusually large inclination between magnetic axis and rotation axis.

DYNAMO PROBLEM

M.M. Vishik
Soviet Geophysical Committee
Molodezhnaya 3, Moscow, USSR

1. To solve the kinematic dynamo problem in case when there is no inner scale.
1.1 What is the asymptotical behavior of the spectrum of the problem

$$\bar{\nabla} \times (u \times H) + \nu \Delta H = \sigma H, \quad (\bar{\nabla}, H) = 0, \quad \nu \to 0?$$

under suitable boundary conditions? Here u is a solenoidal vector field of general type with zero normal component on the boundary.

1.2 What is the connection of the asymptotics, mentioned in 1.1 with the asymptotical behavior of the dynamical system

$$\dot{x} = u(x)?$$

1.3 What is the connection of both asymptotics with quasiclassics for Shroedinger operator and with Fourier integral operators with complex phase?

1.4 What is the asymptotical behavior of Green's function for induction equation

$$\dot{H} = \bar{\nabla} \times (u \times H) + \nu \Delta H, \quad (\bar{\nabla}, H) = 0$$

while $t \to \infty$, $\nu \to 0$?

2. Give the model for changing of the dipole polarity for the magnetic field of the Earth, following from the full MHD system. What is the effective dimension of the attractor in this problem?

POLARITY TRANSITIONS

Kenneth L. Verosub
Department of Geology
University of California, Davis

Even after twenty years of research, polarity transitions of the Earth's magnetic field represent one of the most important and least understood of geophysical phenomena. Key questions to be addressed in the coming years are:

1. What are the characteristics of polarity transitions as seen at single sites, particularly sites in the Southern Hemisphere?
2. What is the global pattern of directional changes for a given polarity transition?
3. What are the characteristics of successive transitions as seen at a given site?
4. What is the relationship between the duration of the intensity change and that of the directional changes?
5. How rapidly do the changes in direction and intensity occur?
6. How long does it take for the field to reverse?
7. What is the nature of secular variation before or after a transition?

8. Are transitions preceded or followed by geomagnetic excursions?
9. Are geomagnetic excursions aborted reversals?
10. Are there phenomenological models which provide an adequate description of the behavior of the field during a transition?
11. Can the phenomenological models be made more realistic by incorporating information about the behavior of the field during the last 400 years?
12. Is there a periodicity in the frequency of transitions?
13. Can the phenomenological models be used to generate time series with the same statistical properties as the Magnetic Polarity Time Scale?
14. Can polarity transitions be analyzed in terms of chaotic processes?
15. Is there a theoretical basis to the phenomenological modelling?

GEOMAGNETISM

V.I. Pochtarev
Leningrad Department of the Institute of Terrestrial Magnetism, Ionosphere and Radio Waves Propagation
Leningrad, USSR

Key Questions

1. What is the nature of the geomagnetic field and its secular variations?
2. What is the mechanism of magnetism of the Sun and of the planets?
3. What is the nature of the large regional magnetic anomalies discovered on the Earth not long ago?
4. What is the contribution of external sources in geomagnetic field and secular variations?
5. What is the relationship between geomagnetic, gravimetrical and other geophysical fields?

GEOMAGNETIC FIELD REVERSALS

S. Stefanescu
Romanian National Committee of Geodesy and Geophysics
R - 70201 Bucharest - 37, Romania

Worldwide reliable assertion of the reality of minor peculiarities of the phenomenon (events, excursions) and elaboration of a mechanism likely to produce it. Some light might be expected in this respect from taking into account the existence of open magnetic field lines and their "catastrophic" behavior when the generating current system -- in the case of the terrestrial self-exciting dynamo: the equatorial electric currents in the Earth's core -- deviates from a planar distribution.

SOLAR TERRESTRIAL PHYSICS

M.I. Pudovkin
Institute of Physics,
Leningrad University, Leningrad, USSR

Three problems are:

1. To establish the resonance phenomena in the Solar System; the existence (or non-existence) of "cosmic" periods in geophysical and biological phenomena.
2. To find the origin of long-period variations in the solar activity.
3. To find a physical model of MHD-generator at high latitude magnetopause.

MATHEMATICAL MODELLING OF THE IONOSPHERE

A.A. Namgaladze
Kaliningrad Observatory of the IZMIRAN
USSR Academy of Sciences, Kaliningrad

1. What are the mathematical models of the influence on the ionosphere of the following processes: the weather perturbations, the explosions in the lower and middle atmosphere and on the Earth's surface, the earthquakes, the eruptions of the volcanoes?

IONOSPHERE DYNAMICS

B.N. Gershman
Gorky University, Gorky, USSR

1. Are there small-scale inhomogeneities in the ionospheric plasma (with scales approximated to the Debye radius) and what part do they play in problems of anomalous resistance and anomalous absorption of radio waves in high and mid latitudes?
2. What role do strong small-scale plasma wave-turbulence play on physical processes in the auroral ionosphere?

CLIMATE, FEEDBACK LOOPS

J.S.A. Green
School of Environmental Sciences
University of East Anglia, Norwich, U.K.

By exploiting powerful computers and satellite observations, weather prediction has improved substantially towards an estimated ultimate. Climate prediction is in a comparatively poor state.

Thus any weather prediction model, which has been allowed the freedom to move away from "present climate" does so until it attains a climate that is outrageous. This statement might be queried; how tightly a particular model is constrained to present conditions by "tuning" is not always clear. Substantial resources have been devoted to the climate problem and political issues are never far away; weather is the prerogative of the State whereas climate is an academic exercise.

We can see how well models reproduce a significant parameter. The mean surface wind in middle latitudes is one such number. The models of the 1980's produce an enormous scatter of values, which only just include the climatic mean. Winter-continental anticyclones, whose essential physics has been known since the 1900's, bear little relation to reality.

Climate is fundamentally different from weather. Weather prediction is the result of repeated accurate calculation of the change of variables from given initial values. Climate relies on balance between diverse physical processes, rather than the accurate representation of any. Feedback and equilibria are the essence.

A very simple feedback loop within the dynamics of weather prediction models concerns the relation between the compression of air and the resulting pressure force. This describes the propagation of sound waves, and initially caused considerable difficulty for the integration of the equations, even though the final balanced state (of hydrostatic balance between the pressure and the weight of air) is well known and easy to visualize.

One primitive climate model has (only) Ice-Albedo feedback; ice reflects sunlight, there is more ice if it is cold. It is naive, simple minded, and violently unstable. Thus the ice sheet either melts completely, or takes over the whole planet. This, and many other catastrophes, may well be hidden in the feedback loops of complex models.

Photochemistry illustrates one problem. Ozone is a fairly simple gas, but the routes by which it can be destroyed include reactions with many obscure chemicals. Some of these reactions are catalytic so the agent is not consumed and therefore assumes an inordinate role.

There is "the stomatal gap". Evaporation by plants is an important part of the water budget over land. Botanists point out that this water must have come through the stomata, and deduce that it is essential that we should know the conditions under which they will open, to let the water out. Is this a detail that is avoided when we consider the natural evolution of plant communities, or is it an essential part of the climatology?

The chemistry and photoactivity of the surface layers of the sea depend on life processes. Experts again say that we do not know enough about the field to do "what you want to do."

But here is the saving grace, or the killing sin, for we don't know what it is that we need to know until we try to do something, like determine climate. Feedback loops have insensitive areas as well as sensitive, and my problem is the finding of critical pathways through large sets of interactive processes. Catastrophe theory may help, but I suspect that we need a new kind of polymath to navigate us through the morass of current scientific detail and elaboration to some overall goal.

ATMOSPHERIC DYNAMICS, CLIMATE AND THEIR PREDICTABILITY

D.M. Sonechkin
Hydrometeorological Research Center of the USSR
Moscow, USSR

A. Introduction

Atmospheric general circulation (AGC) models with complicated stochastic behavior can be considered as dynamical systems with strange attractors.

It is appropriate to characterize the climate as an invariant probability measure concentrated on this strange attractor. In the case of existing several attractors abrupt changes in the current climate and climatic hysteresis are likely to take place.

There are all frequencies in the temporal oscillation spectrum of stochastic dynamical systems, i.e. in principle there may exist weather anomalies of any periods without changing of external forcings.

B. The first group of key questions

1.1 What is the AGC attractor in models and in reality? Is the attractor a single one?
1.2 What is the dimension of the AGC attractor? Does this dimension depend upon motion scales under consideration? I.e. is the AGC attractor a non-homogeneous fractal?
1.3 What is the sensitivity of the AGC in models and in the real atmosphere to external forcing changes determined by? Does this sensitivity represent the change of the dimension of the attractor and its probability measure?
1.4 What equations are describing the evolution of the climatic probability measure when the change of external forcings on the atmosphere occurs? (e.g. when the change of the ocean heat state takes place).

C. The second group of key questions

2.1 What manner does the predictability dependence upon the atmosphere scale motion? Is this predictability determined by the dimension of an attractor in this scale?
2.2 What is the exciting mechanism of interannual variations of the atmospheric general circulation? What role do the super- and subharmonic resonances play between the frequency of the annual variations of the heating to the atmosphere and the characteristic eigen frequencies of variations of the atmospheric general circulation?
2.3 What are the exciting and maintaining mechanisms of processes as blocking one? What role do the phase velocity resonances of planetary waves play in blocking dynamics?
2.4 Is it reasonable to use the inverse value of the characteristic time of air particle running around the Earth? And/or is it reasonable to use the viscous and heat relaxation time as a small parameter in order to separate the synoptic scale motions from the planetary scale motions?

2.5 Whether the moving frame of reference, or the frame with amplitudes and phases of waves as coordinates is more convenient in the planetary motion modelling and long-term weather forecasting?

GRAVITY ANOMALIES AND CLIMATE

E.P. Borisenkov
A.I. Voeikov Main Geophysical Observatory
Leningrad, USSR

1. What are the physical mechanisms of the effect of gravity anomalies caused by the Earth's shape and its internal structure on atmospheric circulation and cyclogenesis, in particular the tropical zone cyclogenesis? To what extent do experimental data corroborate the relationship between gravity anomalies and atmospheric circulation and cyclogenesis?
2. In what way can the effect of gravity anomalies be included in the equations of geophysical hydrodynamics and in the problems of modelling the general circulation and tropical cyclones?

THE PROBLEMS OF ATMOSPHERIC OZONE STUDIES

A.Kh. Khrguian
Physics Faculty
Moscow State University, USSR

1. What is the role of different trace gases, and primarily ozone, which interact in the atmosphere with the participation of solar and space radiation and in this way form new types of particles?
2. What is the participation of biogenic (N_2O, CH_3Cl), anthropogenic (SO_2 etc.), volcanogenic (for example, COS) atmospheric particles in the photochemical atmospheric processes? What are their sources, cycles, transportation and interaction with ozone? What are the new requirements to their observations?
3. What is the climatic impact of trace gases (in particular ozone and labile anthropogenic particles) in the present and future? What is the possible influence of their changes on thermal structure, dynamics and circulation of atmosphere and on synoptic processes and weather phenomena? What is their influence on polar regions which are particularly sensitive in this aspect?

HYDROLOGY

A. Rango
Hydrology Laboratory, Agriculture Research Service
Beltsville, Maryland

A. Introduction

The major problems in hydrology can be delineated by considering both the immediacy of the problem and the length of time needed for completing the necessary research. Time scales of near term (0-15 years), medium term (5-20 years), and long term (15-40 years) and most important associated problems are thus defined.

B. Key Problems

1. Near Term
How can the behavior of organic and inorganic chemicals as they move through the root, vadose, and groundwater zones be better understood in order to predict their occurrence and concentration?
   Improved understanding and quantification of water and pollutant transport is needed to prevent contamination of groundwater supplies. Existing technologies do not provide adequate methods of sampling, measuring, or characterizing the spatial and temporal variability of water and chemical movement. New measurement techniques must be developed to acquire data for a variety of new models that can be used to predict the impact of alternative agricultural chemical applications, water management techniques, and production systems on groundwater quality.
2. Medium Term
How can new technological advances be used to improve hydrological modelling?
   The overall objectives are to increase model understanding, realism, and ease of application, especially in ungauged basins. Models need to be simplified (with utilization of physically-based parameters) and implemented on microcomputers to make them accessible. To facilitate application by inexperienced users, expert systems need to be developed employing artificial intelligence techniques. Remote sensing should be emphasized to encourage real measurement of hydrological variables instead of reliance on point measurements. Remote sensing would be useful for monitoring state variables such as soil moisture, snow water equivalent, precipitation, temperature, and land cover. This will require development of new deterministic models based on real data. Utilization of geographical information systems and real-time data relay with other technological advances would be required for hydrological forecasting.
3. Long Term
What effects do large scale hydrological processes have on food production, weather and climate, and worldwide water supply?
   In order to evaluate these important effects,

hydrological modelling will place increasing emphasis on regional, continental, and global processes. The measurement focus will be shifting, as an example, away from the individual trees to large vegetation units. Development of a permanent Earth observations space facility around the year 2000 will be a key factor in permitting large area hydrological modelling to be conducted successfully.

INTEGRATED FORMULATION OF HYDROLOGICAL MODELS AND DESIGN OF HYDROLOGICAL MONITORING SYSTEMS

L. Ubertini
L. Natale
Dipartimento di Ingegneria Idraulica e Ambientale
Facoltà di Ingegneria, I-27100 Pavia

The relatively recent development of hydrology as an autonomous discipline has been achieved based on other applied sciences and developed only in small part on its own theoretical and experimental base.

Even using the most up-to-date techniques to acquire and to transmit experimental data, measurement systems have nearly always been arranged in order to monitor input and output quantities of the phenomena with only a marginal attention to the knowledge of the hydrological system status. The tendency to use in hydrology an output modelling was favored, between 1960 and the first half of the 1970s, through the use of the techniques of analysis of the system theory and the applications of time series statistics.

This tendency was widely justified by the necessity to obtain rapid results for practical use since hydrology appeared from the beginning as an applied discipline rather than a speculative science.

Hydrological quantities are now more extensively monitored (in space and time) through phenomena of distributed parameters based on classic mechanics equations of the movement of continuous media and adapting them to physical reality, often not well known as opposed to the earlier simulating schemes of concentrated parameters (conceptual models, black-box models).

Continuity conditions as related to the hydrological phenomenon, though needed for the validity of classical equation, are seldom valid in practice.

In particular, difficulties arise in deriving boundary conditions for sudden discontinuities with respect to state changes both in time and space. A solution is usually attained by assuming a continuous medium: by this way a solution to the equations can be obtained, which is useful in practice; however the description of the phenomenon has no local validity regarding discontinuities. In their final form, parameter distributed models appear quite similar to parameter concentrated models. A classical example in surface hydrology consists in the interaction between surface runoff, inter-flow and base-flow on hill slopes during floods.

Analogically the propagation law for a steepfront flood wave moving on a dry surface has still to be properly defined.

Two types of problems are therefore connected with the development of mathematical models in hydrology.

In hydrology, when formulating mathematical models, new interpretative schemes should be employed in order to apply equations derived from fluid mechanics, since the hydrological phenomena are only partially described in terms of classical hydraulics.

Very often, the mathematical problem itself, which comes from the formulation of the hydrological model, changes with the simplifications being introduced and with the numerical values of the equation parameters as well. In some cases (e.g. monodimensional flood routing), both the logical framework of the model and its physical meaning are already well known; but sometimes the output from the mathematical model has to be controlled with regard to the physical response of the natural phenomenon. This is the case of an approximate mathematical description using only continuity equations together with an empirical equation of motion or a simplified state equation. When designing monitoring systems, it would be advisable to insert the measurement of state parameters by giving a detailed description of the hydrological phenomenon thus helping in formulating a mathematical model.

The experimental investigations should be extended to hydrology in a systematic way.

In fact, the experimental investigation performed through laboratory experiences is at the basis of all mathematical theories in hydraulics, but for field surveys it has not yet been applied because of its high cost and duration.

HYDROLOGY AS A CONTRIBUTION TO WATER MANAGEMENT MARKED BY A HIGH SENSE OF RESPONSIBILITY

P. Krausneker
Federal Ministry of Agriculture and Forestry
Hydrographical Central Bureau
A-1190 Vienna, Austria

Introduction

The ancient auxiliary science of hydraulics and the more recently established science of hydrology -- both concentrating on quantities of water -- are highly developed instruments indispensable for water management. The endeavors of these sciences to find optimum solutions have progressed to such an extent that the "objective" decisions (with regard to water management and economic aspects) are interwoven with motives based on political reasoning.

After the old objectives -- protection from the forces of nature and utilization of the natural resources -- a new aspect has come to the fore: Protection of natural water resources from the effects of man's economic activities. Here considerable progress has been made with regard to solving this problem by taking local measures within the water cycle. Nevertheless, great efforts will yet be required to correct the mistakes made during the last few decades, which specially endanger subterranean water resources. As far as overall measures (ground, plants) are concerned, they will only lead to the control of dangerous influences after the priorities for some human activities have been newly formulated. The considerations so far mentioned only apply to our highly industrialized country whose contributions to solving water management problems in the so-called Third World -- and thus to the solution of the North-South conflict -- are hardly worth mentioning. This applies both to the macro-economic relations (e.g. transfer of technology and capital) and the micro-economic aspects of our behavior as consumers. (Which consequences are drawn from the statistics based in US investigations quoted by Van Dam, according to which the specific water requirements for producing 1 kg of beans containing protein amount to 10 liters, whereas those for the production of 1 kg of first-class beef are as high as 30,000 liters?)

Problems

For the above reasons, the fields of problems to be solved by hydrology in future will also extend to "non-technical" (in the conventional sense of the word) questions such as the following:

1. What is the element water? With regard to this question, e.g. Schwenk impressively described the "vital formative capacity" of water in a book which, though utilized by his own institute, receives less attention in connection with classical theory than practical water management.
1.1 Which terrestrial and cosmic factors influence the above-mentioned "vital formative capacity"?
1.2 How can such an influence be proved statistically and how can it be explained deterministically?
1.3 Which conclusions may be drawn from the findings in the interest of the optimum utilization of water?
2. Who am I, also in my capacity as hydrologist?
2.1 How do I deal with the interaction between the "active" subject and the "passive" object in the relationship man-water (an interaction which cannot be denied even outside alchemistic schools)?
2.2 Have active hydrologists certain characteristics in common that are significantly different to those of other people?

The study of this question is essential in order to be in a position to deal with the following question:

3. How is water management actually carried out? What is the interaction between water management and social systems? A contribution to this question was made in the book on the political history of water management written by Wittfogel which has long been a classic in the sciences related to society (the term "hydraulic society" having become a well-established technical term in these circles) even though this book is hardly read by persons active in water management.
3.1 Which influence is exerted on hydrological research by the manner in which a society is organized?
3.2 How does any type of society influence the application of acquired knowledge in connection with the implementation of water management projects?
3.3 Which part is played in this context by the civilizational parameters e.g. by the extent to which a society is industrialized?

Summary

From the great age of pioneers of water management which -- as we learned at school -- began with the advanced civilizations at the rivers Euphrates and Nile, there emerged hydrology -- a thriving field of research activities -- which was largely considered on its own without taking into account the forces influencing it and which it influences. The questions mentioned endeavor to throw a light on the direction in which this discipline will develop in future: Interwoven with important spheres of influence, hydrology will develop into a branch showing a great deal of consideration for human beings and a high sense of responsibility towards nature.

Bibliography

Van Dam, A. : Die Wasserkrise, Entwicklung und Zusammenarbeit Nr. 6/1977, Deutsche Stiftung für Internationale Entwicklung, Bonn, 1977.
Schwenk, T.: Das sensible Chaos. Verlag Freies Geistesleben, Stuttgart, 1962.
Wittfogel, K.: Oriental Despotism. 1957. German: Die orientalische Despotie. Ullstein, Frankfurt, 1977.

LONG-RANGE RISK ANALYSIS:
UNCERTAINTY AND FUZZINESS

L. Duckstein
Department of Systems and Industrial Engineering
University of Arizona, Tucson
H.P. Nachtnebel
Department of Water Resources Engineering
Unversität für Bodenkultur, A-1180 Wien
I. Bogardi
Civil Engineering Department,
University of Nebraska, Lincoln

A. Introduction

Risk analysis includes the definition of scenarios, the estimation of the probability of a given scenario and the assessment of the consequences stemming from the occurrence of an adverse event or scenario. These consequences may be factual or perceived and are usually described from the viewpoint of multiple criteria. In intermediate or long-range problems, randomness, uncertainty and imprecision (or vagueness) may be present.

B. Key Questions

1. Which elements are random, uncertain, or fuzzy?
2. How can membership functions be assessed reliably and then combined?
3. How can a fuzzy set membership function best be used

as an indicator of the effect of imprecise environmental consequences?
4. How can the trade-offs be made to manage the risk?

RHEOLOGY OF METAMORPHIC ROCK ICE

L. Lliboutry
University of Grenoble I, and Laboratoire
de Glaciologie et Géophysique de l'Environnement du CNRS
38402 Grenoble, France

A. Introduction

Although polycrystalline ice is, after basalt, the most widespread rock on the Earth's surface, it is generally ignored by investigators dealing with rock rheology or thermal dynamo-metamorphism. However, processes which are out of reach in the laboratory could be thoroughly studied with polycrystalline ice, which deforms and evolves, in natural conditions, about 10^6 times faster. Moreover, its varying petrography can be easily studied by optical means.

Coring polar ice caps and ice shelves has revealed the existence of five or six distinct kinds of metamorphic rock ice, with different rheologies. The creep that can be considered as permanent at time scales shorter than 1000 years (otherwise the slow change in fabrics has to be introduced) may be either superplasticity linked with grain growth, or Newtonian dislocation creep, or power-law dislocation creep (either isotropic or anisotropic), or dislocation recrystallization creep (either isotropic or anisotropic). Among these kinds of ice, several ones cannot be reproduced in the laboratory. Moreover, during creep tests, transient creep is observed during the first days.

Besides the necessity of having correct constitutive laws for modelling realistically the flow of actual ice sheets, and their waxing and waning in the past, ice studies could lead to deep insights on pending problems as the viscosity of the lower mantle (the one pertinent to convection, and the one pertinent to glacio-isostatic rebound), or the first stages of magma migration.

B. Key Questions

1. Reasons why rock ice formed during the last Glacial has much smaller grain size and is much more anisotropic than rock ice formed during Holocene.
   An explanation might lead to the knowledge of another aspect of past climates than stable isotope studies (which yield the mean temperature above the inversion layer, and the total ice volume on Earth).
2. Rheological parameters of anisotropic ice as a function of the c-axes distribution.
   Actual homogenization thoeries do not fit the data, probably because grain boundary migration is ignored.
3. Factors governing the onset of recrystallization in anisotropic rock ice.
   Then, a peculiar four-maxima fabric is formed, and the shear strain rate becomes lower. This fabric might be related to the formation of coincidence site lattices, and thus is a problem in fundamental science.
   However, it is also essential for correctly modelling the flow or a polar ice sheet.
4. General constitutive law for transient creep.
   In uni-axial creep tests a recoverable, logarithmic strain (akin to delayed elasticity), and a permanent strain which follows Andrade's law can be distinguished. To model transient flows, the complete strain rate tensor due to a varying stress deviator must be known.

AUTHOR INDEX

Author	Title	Page
Aki, K	Earthquake Research	104
Akimoto, S	The Structure and Dynamics of the Earth's Deep Interior	94
Banks, RJ	Electrical Conductivity	106
Baran, W	Space Geodesy	90
Beloussov, VV	Continental and Oceanographic Lithospheres	101
Biró, P	Gravity Field Variation	91
Biró, P	Geoid Interpretation	94
Bogardi, I	Long-Range Risk Analysis: Uncertainty and Fuzziness	112
Borisenkov, EP	Gravity Anomalies and Climate	110
Busse, FH	Geodynamo	106
Copaciu, CC	Earth Mass—Stability and Fluctuations	85
Crampin, S	Cracks in the Crust: Evidence and Implications	100
Crutzen, PJ	Comments on George Reid's "Quo Vadimus" Contribution "Climate"	47
Dobrovolski, SG	Global Water and Heat Exchange	87
Dobrovolsky, IP	Lithosphere Dynamics	99
Duckstein, L	Long-Range Risk Analysis: Uncertainty and Fuzziness	112
Eagleson, PS	Statement on Quo Vadimus-Hydrology	79
Emelyanov, AP	Earthquake Research	104
Evison, FF	Seismogenesis, Prediction of Earthquakes and Mitigation of Earthquake Losses	103
Gabrielov, AM	Lithosphere Dynamics	97
Gao, L-S	Intraplate Earthquakes	103
Garetsky, RG	Earthquake Research	104
Gershman, BN	Ionosphere Dynamics	108
Grafarend, EW	Geodesy	89
Green, JSA	Climate, Feedback Loops	108
Guangzhi, T	Multiple Sources and Multigenesis, the Problem Facing Geochemistry, Petrology, and Metallogeny—with Special Reference on Skarns and Skarn Deposits in China	105
Harrison, CGA	The Core of the Earth	95
Hide, R	Geophysical Fluid Dynamics and Related Topics	39
Holota, P	Physical Geodesy	90
Kamide, Y	The Importance of the Variability of the Solar-Terrestrial Environment	23
Kaula, WM	The Earth as a Planet	13
Kautzleben, H	Strain and Stress of the Earth Crust	99
Keilis-Borok, V	The Lithosphere of the Earth as a Large Non-Linear System	81
Khamrabaev, I.Kh.	Crust and Mantle	96
Kharadze, EK	Solar Terrestrial Connections; Earthquake Precursors; Atmospheric Aerosol	85
Khrguian, A. Kh.	The Problems of Atmospheric Ozone Studies	110
Krausneker, P	Hydrology as a Contribution to Water Management Marked by a High Sense of Responsibility	111
Kropotkin, PN	Mantle Dynamics	96
Kundzewicz, ZW	Quo Vadimus—Hydrology	71
Kuo, JT	Atmosphere-Earth-Ocean Dynamics	87
Kurlenja, MB	Physics of Solid Rocks	104
Lal, D	Oceanography and Geophysics	59
Lambeck, K	Geophysical Geodesy: The Study of the Slow Deformations of the Earth	7
LeBlond, PH	Earth Science in the 21st Century	63
Lliboutry, L	Rheology of Metamorphic Rock Ice	113
Matthews, DH	The Rheology of the Lithosphere	98
McCarthy, DD	Earth Orientation and Reference Systems—The Future	92
Meier, M	Statement to Follow Quo Vadimus—Hydrology by Z.W. Kundzewicz	77
Monin, AS	Mantle Convection	96
Moritz, H	Geodesy and Geophysics in Their Interaction with Mathematics and Physics, and Some Open Problems in Geodesy	1
Mustel, ER	Earth Rotation	93
Mutajwaa, AL	Development of Geodesy and Geophysics in Africa (East Africa Case)	88
Nachtnebel, HP	Long-Range Risk Analysis: Uncertainty and Fuzziness	112
Nagata, T	The Structure and Dynamics of the Earth's Deep Interior	94
Namgaladze, AA	Mathematical Modeling of the Ionosphere	108
Namias, J	Where are We Going in the Study of Short-Period Climate Fluctuations?	25
Natale, L	Integrated Formulation of Hydrlogical Models and Design of Hydrological Monitoring Systems	111
Oparin, VN	Physics of Solid Rocks	104
Peltier, RW	Hydrodynamic Complexity in the Earth System	43
Pochtarev, VI	Geomagnetism	108
Pudovkin, MI	Solar-Terrestrial Physics	108
Rango, A	Hydrology	110
Reid, GC	Climate	45
Runcorn, SK	Gravity Field of the Earth and the Dynamics of its Interior	93
Sarkisov, YM	Asthenolayer of the Continental Crust, Facts and Problems	99
Schneider, SH	Research in Climate Science	49
Shemiakin, EI	Physics of Solid Rocks	104
Siscoe, GL	Space Plasma Physics	21
Snow, JT	Scale Interactions	55
Sonechkin, DM	Atmospheric Dynamics, Climate and Their Predictability	109
Stefanescu, S	Geodesy and Gravity	89
Stefanescu, S	Geomagnetic Field Reversal	108
Stewart, RW	Physical Oceanography to the End of the Twentieth Century	65
Torge, W	Gravity Field	91
Treder, H.-J.	Experimental Gravity Research	89
Troitskiy, VS	The Geoid	90
Trubizin, VP	Mantle Structure	96
Ubertini, L	Integrated Formulation of Hydrological Models and Design of Hydrological Monitoring Systems	111
Uyeda, S	Geodynamics of Subduction	97
Vaniček, P	Some Possible Additional Answers	11
Verosub, KL	Polarity Transitions	107
Vishik, MM	Dynamo Problem	107
Vlaar, NJ	Vening Meinesz, a Pioneer in Earth Sciences	xi
Volvovsky, BS	Asthenolayer of the Continental Crust, Facts and Problems	99
Volvovsky, IS	Asthenolayer of the Continental Crust, Facts and Problems	99
Wahr, JM	Geodesy and Geophysics	5
Wilkins, GA	The Rotation of the Earth	93
Williams, DJ	Space Plasma Physics	21
Wilson, L	Physical Volcanology	106
Wunsch, C	Comments on R.W. Stewart's "Physical Oceanography to the End of the Twentieth Century"	69
Zielinski, JB	Earth Size Changes	91

SUBJECT INDEX

ATMOSPHERIC COMPOSITION AND STRUCTURE

Crutzen, PJ   Comments on George Reid's "Quo Vadimus" Contribution "Climate"   47
Schneider, SH   Research in Climate Science   49
Kharadze, EK   Solar Terrestrial Connections; Eathquake Precursors; Atmospheric Aerosol   85
Green, JSA   Climate, Feedback Loops   108
Sonechkin, DM   Atmospheric Dynamics, Climate and Their Predictability   109
Khrguian, A. Kh.   The Problems of Atmospheric Ozone Studies   110

ELECTROMAGNETICS

Moritz, H   Geodesy and Geophysics in Their Interaction with Mathematics and Physics, and Some Open Problems in Geodesy   1
Harrison, CGA   The Core of the Earth   95
Monin, AS   Mantle Convection   96
Banks, RJ   Electrical Conductivity   106
Busse, FH   Geodynamo   106
Vishik, MM   Dynamo Problem   107
Verosub, KL   Polarity Transitions   107
Stefanescu, S   Geomagnetic Field Reversal   108

EXPLORATION GEOPHYSICS

Geodesy, Geophysics and Remote Sensing   88
Crampin, S   Cracks in the Crust: Evidence and Implications   100
Guangzhi, T   Multiple Sources and Multigenesis, the Problem Facing Geochemistry, Petrology, and Metallogeny—with Special Reference on Skarns and Skarn Deposits in China   105

GEODESY AND GRAVITY

Vlaar, NJ   Vening Meinesz, a Pioneer in Earth Sciences   xi
Moritz, H   Geodesy and Geophysics in Their Interaction with Mathematics and Physics, and Some Open Problems in Geodesy   1
Wahr, JM   Geodesy and Geophysics   5
Lambeck, K   Geophysical Geodesy: The Study of the Slow Deformations of the Earth   7
Vaniček, P   Some Possible Additional Answers   11
Hide, R   Geophysical Fluid Dynamics and Related Topics   39
Peltier, RW   Hydrodynamic Complexity in the Earth System   43
Copaciu, CC   Earth Mass—Stability and Fluctuations   85
Kuo, JT   Atmosphere-Earth-Ocean Dynamics   87
Mutajwaa, AL   Development of Geodesy and Geophysics in Africa (East Africa Case)   88
Geodesy, Geophysics and Remote Sensing   88
Treder, H.-J.   Experimental Gravity Research   89
Grafarend, EW   Geodesy   89
Stefanescu, S   Geodesy and Gravity   89
Holota, P   Physical Geodesy   90
Baran, W   Space Geodesy   90
Troitskiy, VS   The Geoid   90
Zielinski, JB   Earth Size Changes   91
Torge, W   Gravity Field   91
Biró, P   Gravity Field Variation   91
McCarthy, DD   Earth Orientation and Reference Systems—The Future   92
Mustel, ER   Earth Rotation   93
Runcorn, SK   Gravity Field of the Earth and the Dynamics of its Interior   93
Wilkins, GA   The Rotation of the Earth   93
Biró, P   Geoid Interpretation   94
Harrison, CGA   The Core of the Earth   95
Monin, AS   Mantle Convection   96
Kropotkin, PN   Mantle Dynamics   96
Trubizin, VP   Mantle Structure   96
Kautzleben, H   Strain and Stress of the Earth Crust   99
Borisenkov, EP   Gravity Anomalies and Climate   110

GEOMAGNETISM AND PALEOMAGNETISM

Hide, R   Geophysical Fluid Dynamics and Related Topics   39
Peltier, RW   Hydrodynamic Complexity in the Earth System   43
Copaciu, CC   Earth Mass—Stability and Fluctuations   85
Harrison, CGA   The Core of the Earth   95
Banks, RJ   Electrical Conductivity   106
Busse, FH   Geodynamo   106
Vishik, MM   Dynamo Problem   107
Verosub, KL   Polarity Transitions   107
Stefanescu, S   Geomagnetic Field Reversal   108
Pochtarev, VI   Geomagnetism   108

HISTORY OF GEOPHYSICS

Vlaar, NJ   Vening Meinesz, a Pioneer in Earth Sciences   xi
Moritz, H   Geodesy and Geophysics in Their Interaction with Mathematics and Physics, and Some Open Problems in Geodesy   1
LeBlond, PH   Earth Science in the 21st Century   63
Kundzewicz, ZW   Quo Vadimus—Hydrology   71
Beloussov, VV   Continental and Oceanographic Lithospheres   101

HYDROLOGY

Kundzewicz, ZW   Quo Vadimus—Hydrology   71
Meier, M   Statement to Follow "Quo Vadimus--Hydrology" by Z.W. Kundzewicz   77
Eagleson, PS   Statement on "Quo Vadimus-Hydrology"   79
Dobrovolski, SG   Global Water and Heat Exchange   87
Green, JSA   Climate, Feedback Loops   108
Rango, A   Hydrology   110
Krausneker, P   Hydrology as a Contribution to Water Management Marked by a High Sense of Responsibility   111
Ubertini, L   Integrated Formulation of Hydrolog-

ical Models and Design of Hydrological Monitoring Systems	111
Natale, L Integrated Formulation of Hydrlogical Models and Design of Hydrological Monitoring Systems	111
Lliboutry, L Rheology of Metamorphic Rock Ice	113

IONOSPHERE

Kamide, Y The Importance of the Variability of the Solar-Terrestrial Environment	23
Gershman, BN Ionosphere Dynamics	108
Namgaladze, AA Mathematical Modeling of the Ionosphere	108

MAGNETOSPHERIC PHYSICS

Siscoe, GL Space Plasma Physics	21
Williams, DJ Space Plasma Physics	21
Kamide, Y The Importance of the Variability of the Solar-Terrestrial Environment	23
Pudovkin, MI Solar-Terrestrial Physics	108

MARINE GEOLOGY AND GEOPHYSICS

Vlaar, NJ Vening Meinesz, a Pioneer in Earth Sciences	xi

METEOROLOGY AND ATMOSPHERIC DYNAMICS

Namias, J Where are We Going in the Study of Short-Period Climate Fluctuations?	25
Hide, R Geophysical Fluid Dynamics and Related Topics	39
Peltier, RW Hydrodynamic Complexity in the Earth System	43
Reid, GC Climate	45
Schneider, SH Research in Climate Science	49
Snow, JT Scale Interactions	55
Lal, D Oceanography and Geophysics	59
LeBlond, PH Earth Science in the 21st Century	63
Eagleson, PS Statement on "Quo Vadimus-Hydrology"	79
Dobrovolski, SG Global Water and Heat Exchange	87
Green, JSA Climate, Feedback Loops	108
Namgaladze, AA Mathematical Modeling of the Ionosphere	108
Sonechkin, DM Atmospheric Dynamics, Climate and Their Predictability	109
Borisenkov, EP Gravity Anomalies and Climate	110
Rango, A Hydrology	110
Khrguian, A. Kh. The Problems of Atmospheric Ozone Studies	110

MINERALOGY, PETROLOGY, AND ROCK CHEMISTRY

Copaciu, CC Earth Mass—Stability and Fluctuations	85
Akimoto, S The Structure and Dynamics of the Earth's Deep Interior	94
Nagata, T The Structure and Dynamics of the Earth's Deep Interior	94
Khamrabaev, I.Kh. Crust and Mantle	96
Trubizin, VP Mantle Structure	96
Uyeda, S Geodynamics of Subduction	97
Gabrielov, AM Lithosphere Dynamics	97
Matthews, DH The Rheology of the Lithosphere	98
Guangzhi, T Multiple Sources and Multigenesis, the Problem Facing Geochemistry, Petrology, and Metallogeny—with Special Reference on Skarns and Skarn Deposits in China	105

MINERAL PHYSICS

Lliboutry, L Rheology of Metamorphic Rock Ice	113

OCEANOGRAPHY: GENERAL

Vlaar, NJ Vening Meinesz, a Pioneer in Earth Sciences	xi
Peltier, RW Hydrodynamic Complexity in the Earth System	43
LeBlond, PH Earth Science in the 21st Century	63
Stewart, RW Physical Oceanography to the End of the Twentieth Century	65
Wunsch, C Comments on R.W. Stewart's "Physical Oceanography to the End of the Twentieth Century"	69
Dobrovolski, SG Global Water and Heat Exchange	87
Sonechkin, DM Atmospheric Dynamics, Climate and Their Predictability	109

OCEANOGRAPHY: BIOLOGICAL AND CHEMICAL

Lal, D Oceanography and Geophysics	59
Green, JSA Climate, Feedback Loops	108

PHYSICAL PROPERTIES OF ROCKS

Gabrielov, AM Lithosphere Dynamics	97
Matthews, DH The Rheology of the Lithosphere	98
Dobrovolsky, IP Lithosphere Dynamics	99
Crampin, S Cracks in the Crust: Evidence and Implications	100
Aki, K Earthquake Research	104
Kurlenja, MB Physics of Solid Rocks	104
Oparin, VN Physics of Solid Rocks	104
Shemiakin, EI Physics of Solid Rocks	104

PLANETOLOGY: SOLID SURFACE PLANETS AND SATELLITES

Kaula, WM The Earth as a Planet	13
Geodesy, Geophysics and Remote Sensing	88
Akimoto, S The Structure and Dynamics of the Earth's Deep Interior	94
Nagata, T The Structure and Dynamics of the Earth's Deep Interior	94
Busse, FH Geodynamo	106
Pochtarev, VI Geomagnetism	108

PLANETOLOGY: FLUID PLANETS

Hide, R Geophysical Fluid Dynamics and Related Topics	39

Busse, FH Geodynamo	106
Pochtarev, VI Geomagnetism	108

POLICY SCIENCES

LeBlond, PH Earth Science in the 21st Century	63
Mutajwaa, AL Development of Geodesy and Geophysics in Africa (East Africa Case)	88
Evison, FF Seismogenesis, Prediction of Earthquakes and Mitigation of Earthquake Losses	103
Krausneker, P Hydrology as a Contribution to Water Management Marked by a High Sense of Responsibility	111
Bogardi, I Long-Range Risk Analysis: Uncertainty and Fuzziness	112
Duckstein, L Long-Range Risk Analysis: Uncertainty and Fuzziness	112
Nachtnebel, HP Long-Range Risk Analysis: Uncertainty and Fuzziness	112

RADIO SCIENCE

Gershman, BN Ionosphere Dynamics	108

SEISMOLOGY

Lambeck, K Geophysical Geodesy: The Study of the Slow Deformations of the Earth	7
Keilis-Borok, V The Lithosphere of the Earth as a Large Non-Linear System	81
Runcorn, SK Gravity Field of the Earth and the Dynamics of its Interior	93
Harrison, CGA The Core of the Earth	95
Khamrabaev, I.Kh. Crust and Mantle	96
Monin, AS Mantle Convection	96
Kropotkin, PN Mantle Dynamics	96
Uyeda, S Geodynamics of Subduction	97
Gabrielov, AM Lithosphere Dynamics	97
Matthews, DH The Rheology of the Lithosphere	98
Sarkisov, YM Asthenolayer of the Continental Crust, Facts and Problems	99
Volvovsky, BS Asthenolayer of the Continental Crust, Facts and Problems	99
Volvovsky, IS Asthenolayaer of the Continental Crust, Facts and Problems	99
Dobrovolsky, IP Lithosphere Dynamics	99
Kautzleben, H Strain and Stress of the Earth Crust	99
Crampin, S Cracks in the Crust: Evidence and Implications	100
Gao, L-S Intraplate Earthquakes	103
Evison, FF Seismogenesis, Prediction of Earthquakes and Mitigation of Earthquake Losses	103
Aki, K Earthquake Research	104
Emelyanov, AP Earthquake Research	104
Garetsky, RG Earthquake Research	104
Kurlenja, MB Physics of Solid Rocks	104
Oparin, VN Physics of Solid Rocks	104
Shemiakin, EI Physics of Solid Rocks	104
Banks, RJ Electrical Conductivity	106

SOLAR PHYSICS, ASTROPHYSICS, AND ASTRONOMY

Pochtarev, VI Geomagnetism	108
Pudovkin, MI Solar-Terrestrial Physics	108

SPACE PLASMA PHYSICS

Williams, DJ Space Plasma Physics	21
Siscoe, GL Space Plasma Physics	21
Kamide, Y The Importance of the Variability of the Solar-Terrestrial Environment	23
Gershman, BN Ionosphere Dynamics	108

TECTONOPHYSICS

Lambeck, K Geophysical Geodesy: The Study of the Slow Deformations of the Earth	7
Lal, D Oceanography and Geophysics	59
Keilis-Borok, V The Lithosphere of the Earth as a Large Non-Linear System	81
Copaciu, CC Earth Mass—Stability and Fluctuations	85
Kuo, JT Atmosphere-Earth-Ocean Dynamics	87
Geodesy, Geophysics and Remote Sensing	88
Torge, W Gravity Field	91
Biró, P Gravity Field Variation	91
Runcorn, SK Gravity Field of the Earth and the Dynamics of its Interior	93
Akimoto, S The Structure and Dynamics of the Earth's Deep Interior	94
Nagata, T The Structure and Dynamics of the Earth's Deep Interior	94
Khamrabaev, I.Kh. Crust and Mantle	96
Monin, AS Mantle Convection	96
Kropotkin, PN Mantle Dynamics	96
Trubizin, VP Mantle Structure	96
Uyeda, S Geodynamics of Subduction	97
Gabrielov, AM Lithosphere Dynamics	97
Matthews, DH The Rheology of the Lithosphere	98
Sarkisov, YM Asthenolayer of the Continental Crust, Facts and Problems	99
Volvovsky, BS Asthenolayer of the Continental Crust, Facts and Problems	99
Volvovsky, IS Asthenolayaer of the Continental Crust, Facts and Problems	99
Dobrovolsky, IP Lithosphere Dynamics	99
Kautzleben, H Strain and Stress of the Earth Crust	99
Beloussov, VV Continental and Oceanographic Lithospheres	101
Gao, L-S Intraplate Earthquakes	103
Evison, FF Seismogenesis, Prediction of Earthquakes and Mitigation of Earthquake Losses	103
Emelyanov, AP Earthquake Research	104
Garetsky, RG Earthquake Research	104

VOLCANOLOGY

Wilson, L Physical Volcanology	106
Namgaladze, AA Mathematical Modeling of the Ionosphere	108